Instructor's Manual

to Accompany

Empirical Direction
in Design and Analysis

Instructor's Manual

to Accompany

Empirical Direction
in Design and Analysis

Norman H. Anderson

University of California, San Diego

Routledge
Taylor & Francis Group

NEW YORK AND LONDON

First published by
Lawrence Erlbaum Associates, Inc., Publishers
10 Industrial Avenue
Mahwah, New Jersey 07430

Transferred to Digital Printing 2009 by Routledge
270 Madison Ave, New York NY 10016
2 Park Square, Milton Park, Abingdon, Oxon, OX14 4RN

ISBN 0-8058-4083-4

Publisher's Note
The publisher has gone to great lengths to ensure the quality of this reprint
but points out that some imperfections in the original may be apparent.

CONTENTS

INSTRUCTOR'S MANUAL FOR

EMPIRICAL DIRECTION IN DESIGN AND ANALYSIS

The exercises in *Empirical Direction* aim to be Socratic. They are intended to help students develop research judgment that is sensitive to opportunities and difficulties of each empirical investigation and flexible in dealing with them.

EXERCISES AS FRAMEWORK FOR LEARNING–TEACHING

Learning should seek for transfer; teaching should follow suit. Courses have value only insofar as students learn concepts and techniques that will transfer to their later activities. Exercises are vital to learning as a means to involve students in thinking processes needed in transfer situations.

Research judgment is the prime ingredient of transfer. Each empirical investigation presents its own opportunities and obstacles. Cookbook design/statistics has real value, but it leaves much to be desired. To deal with the uncertainties and problems that arise in any empirical investigation requires educated judgment. The investigator must take account of empirics of each individual situation, consider alternative empirical tasks, measures, and designs, and compare their situation-specific pros and cons. Similar skills are needed to evaluate published articles and other reports. Effective learning–teaching thus requires that statistics be subordinated to and integrated with empirics. This kind of goal has been emphasized by many, but it is not easy to achieve.

Exercises offer an ideal vehicle for development of research judgment. Exercises can be small-scale realizations of problems that arise in actual investigations and of thought processes that can help solve such problems.

DESIRED SKILLS

Several kinds of skills are desirable for design and analysis, and several kinds of exercises are accordingly required. A partial itemization is given here.

Visual inspection is the most important skill for data analysis. Students should know how to assimilate a graph or scan a data set to get some idea of central tendency, pattern, variability, and extreme scores. Visual inspection is vital for understanding empirical significance of data, including substantive size and importance of observed effects. In much research, statistical inference is only a supplement to visual inspection. A number of exercises, accordingly, ask for data analysis based solely on visual inspection.

The value of visual inspection is nicely illustrated with Rayleigh's data on density of nitrogen, the first exercise in this book. What may seem a perceptual dazzle of five-decimal numbers soon reveals a meaningful pattern. This exercise is also a good visual illustration of the qualitative idea of confidence interval and significance test.

The confidence interval is a happy marriage of visual inspection and formal statistics. It represents the mean as an interval of uncertainty and it helps make a substantive judgment of the size of the effect. For understanding pattern of multiple determination, visual inspection of factorial graphs and tables is similarly useful.

Number appreciation is also important, perhaps not really distinct from visual inspection, as with the confidence interval. For the most part, I have used very simple sets of data to help students develop their number appreciation. Not included, I regret to say, are exercises on reading more complicated tables, so well discussed by Wallis and Roberts (1956, pp. 270-279).

Some exercises ask for a modest level of hand calculation, mainly with confidence intervals and power. These exercises aim to bring out algebraic–conceptual structure of the formulas and to give practice in visual inspection, including number appreciation. Hand calculation for one-way analysis of variance is also included, but perhaps this is an outdated holdover from my precomputer days.

Some exercises require justification of some statement in the text or ask the student to role-play a TA in an undergraduate class on research method by explaining some concept to their class. Most psychologists have to process masses of written material; we learn to skim to extract the main point. Skimming is a valuable skill, but statistical concepts and arguments often require concerted thought to gain understanding that will be effective in one's own work. These two kinds of exercises are intended to help students think through what they read.

A general goal of the exercises is to help students subordinate and integrate statistics with extrastatistical empirics. One obstacle to this goal is that much substantive inference is area-specific and task-specific, unfamiliar to many first-year students. Nevertheless, I believe that major components of many statistical–substantive issues can be presented in a general way. Among these are bias from lack of randomization, and confounding from suggestion, personal belief, and other causes (Chapter 8). Other such issues include opportunities and pitfalls of post hoc analysis, problems of extreme scores, and order effects in repeated measures designs.

Reading the literature is a basic skill and some exercises of this type are included. These tend to be fairly long to be at all realistic, so their number is limited. It would be useful if empirical investigators cooperated to collect, develop, and publish collections of such exercises for specific areas.

The concern with research judgment and transfer that appear in the exercises in *Empirical Direction* differs from the cookbook calculations characteristic of other texts. Other texts include answers to selected problems, which is appropriate because students must still get the calculations correct. This goal is achieved here by including answers as part of many of those exercises that require calculation. But with exercises that ask for judgment, giving the answer would short-circuit the mental activity the exercise was designed to elicit. My answers, many of which are presented only as provisional suggestions, are included in this *Instructor's Manual*.

RESEARCH JUDGMENT

The goal of this to book is help its readers develop research judgment. This goal requires an empirical direction, which entails radical reorientation of statistics teaching. This empirical direction should be the organizing principle of texts on design/analysis.

Current texts, however, seem headed in an opposite direction, increasing their emphasis on statistics and losing touch with empirics. Accordingly, I have made strong criticisms of various aspects of current texts.

Such criticism is not new. Many writers have expressed dismay with the current cookbook orientation in statistics teaching. One sign appears in the occasional articles showing that many professional workers are unclear about some elementary statistical ideas. Another sign is the long fusillade against the significance test (Note 19.2a). But these criticisms have had little effect on graduate texts, which suffer from lack of empirical direction (see *Textbook Versus Handbook*, page 769).

Lack of empirical direction may be illustrated with three issues from *Empirical Direction*. One concerns statistical interactions of Anova, which receive adulational treatment in most texts, but which often mean little empirically. The two facts are that Anova interactions are defined as deviations from an arbitrary addition model and that they may be illusions of the response scale. These two facts have been around for decades, but they are virtually ignored in all but one current text (Chapter 7).

A second issue concerns the α escalation that occurs when multiple tests are made. The increasingly hardline advocacy of familywise α in current texts reflects their underlying statistical mentality. In particular, they neglect to mention that many authorities prescribe the opposite practice of using a separate α for each of a family of planned tests. More seriously, this fixation on familywise α loses sight of the continuous, progressive nature of scientific inquiry, especially of the empirical principle of replication (Section 2.4.6 and Chapter 17).

A third issue is analysis with nonrandom groups, especially in field science. Multiple regression is a common technique, standardly presented as a way to "control" or "partial out" systematic preexperimental differences, thereby allowing causal interpretation. This area is a minefield, in which the well-known dangers receive ineffectual lip serivce (Sections 13.2 and 16.2).

These dangers with nonrandom groups are avoided with the randomized experiments used throughout psychology. Being an experimentalist myself, I did not appreciate the importance of field science before writing this book. I believe there are legitimate applications of multiple regression, as suggested in Section 16.2, but they are swamped by the illegitimate. The remedy is to restructure multiple regression on empirical foundations, especially in statistics texts.

Empirical Direction strives for a Socratic emphasis on learning rather than the medieval–theological emphasis on teaching. This empirical direction subordinates statistics to empirics as the foundation for learning research judgment (Chapter 23, *Lifelong Learning*).

TEACHING SUGGESTIONS

Many of the chapter exercises emphasize judgment rather than calculation. These exercises vary in difficulty, allowing each instructor choice to meet needs of different classes.

Answers to these judgment exercises are often not clear-cut, unlike exercises in calculation, and some students may feel uncomfortable with them, especially in the beginning. Preparatory hints and guidance from the instructor on these exercises may facilitate learning.

One tactic is to assign certain exercises on two successive weeks, with an intervening class discussion. The first assignment exposes students to the problem. They can then absorb and benefit from ideas of other students in the class discussion as a base for deeper thinking in the second assignment. Another tactic is to supplement the more open-ended exercises with preliminary pointers to answers, especially early in the course.

I scheduled four class hours weekly, one of which was devoted to the assigned exercises. This exercise hour was mostly student presentation, which generates active participation by other students instead of passive reception of what the instructor says. Assigning a team of two students to be responsible for class presentation of each exercise may be helpful.

To solidify feedback from the class discussion, I allowed students to correct their answers during this exercise hour before handing them in at the end. For this purpose, I suggest that a blank space be required at the end of each answer on their paper in which they can add such corrections and comments. Students liked my written comments on their papers, but I came to feel they did not learn much from this delayed feedback.

I encouraged study groups for doing exercises. Small groups of compatible students may be most effective, and it may be useful to adopt an explicit policy that cooperative work on exercises is desirable and ethical. In this way, exercises can help develop skills of cooperative inquiry for later life.

For use in exams, each student was encouraged to bring a single sheet of formulas or whatever. Making up this sheet helps students organize the material and reduces anxiety. As another anxiety reducer, I generally used a very simple exercise to start off each exam.

I recommend an early exam—a salutary reminder of how slow it can be to assimilate statistical concepts when first exposed to them. I also recommend an announced practice of including past exercises on the exams. This not only motivates students to digest the feedback on the exercises, but provides a sometimes sobering measure of how much they are learning.

Only a handful of statistical concepts is needed, most of which are given in Chapter 2 and expanded in Chapters 3 and 4. The concept of sampling distribution is fundamental, of course, and can be made more intuitive by relating it to the law of sample size. The central limit theorem is equally important, especially as the foundation for confidence intervals. On these concepts is built a superstructure dealing with variability assessment as a means to assess evidence content of a set of data.

Heavy emphasis is placed on confidence intervals as natural measures of response variability and of effect size. These chapters are short but assimilating their content is a continuing process for most of us, particularly in our first graduate year.

Distributed practice seems one of the few known effective ways to facilitate learning. This opportunity can be exploited by spreading some exercises of Chapters 2–4, which cover the basic ideas, over the whole course.

What counts is what students learn. There is substantial agreement that students need to learn to think. Instructors should see themselves as coaches rather than teachers. Except for a gifted minority of instructors, the current lecture system is an inefficient relic from the middle ages. Inevitably, moreover, it imposes a distorted view of education as teaching rather than learning.

Learning is lifelong. To this goal the educational system should dedicate itself. One step in this direction is development of a coaching orientation that emphasizes self-instruction and self-development.

EXERCISES AS RESEARCH DOMAIN

The low level of concern with improving learning effectiveness in most colleges and universities is astonishing. Many instructors are devoted to teaching, of course, and some are good at motivating students to pursue knowledge. A few journals on teaching exist as well as dedicated groups in a few institutions. But overall, the picture is bleak. In psychology, the situation is exemplified by the contrast between, for example, the vast mass of research on rote-reproductive memory and the vast abyss of research on learning–teaching.

From what I have seen, the same faculty-centered perspective is common in other departments. I have been on many interdepartmental committees, which are required for every promotion in the University of California, but in only one did teaching receive more than the perfunctory concern that the regulations require. If colleges and universities took seriously their social responsibility, every department would have at least one senior member devoted to research on science of learning–teaching (see Notes 23.2a,b on page 781 of the text).

Exercises offer one way to break out of the traditional teacher-centered mode to a learner-centered mode of education. Exercises provide an attractive domain for empirical research. Exercises are compact, with many dimensions of variation and other characteristics that make them amenable to empirical analysis. Exercises can embody thinking skills needed for transfer. Good exercises can be cogent tools to investigate the little-known structure of knowledge systems that are operative or effective in lifelong learning.

Exercises should, I believe, be considered scholarly contributions and cooperative knowledge. Development of effective exercises should have high priority in every discipline, humanities included. In statistics, the molehill of unsystematic research on exercises contrasts darkly with the large heaps of research on effects of violations of assumptions of statistical tests. A good exercise should have no less scholarly status than a good research article.

The exercises in *Empirical Direction* are offered as a step toward a science of exercises. I am keenly aware that they have many shortcomings and I welcome advice for improvement:

Norman H. Anderson
Psychology–UCSD
9500 Gilman Drive
La Jolla, CA 92093-0109
nanderson@ucsd.edu

ANSWERS FOR CHAPTER 2

1. By 1890, it was accepted that air was composed solely of oxygen and nitrogen; assiduous chemical investigations had revealed no other chemically active components. To measure density of nitrogen, Lord Rayleigh used hydrogen to burn out the oxygen from a sample containing both oxygen and nitrogen, eliminated the resultant water with a dessicant, and weighed the remainder gas (after overcoming various difficulties). (With thanks to Tukey, 1977, pp. 49*ff.*)

    a. Graph Lord Rayleigh's data in some meaningful way.

    b. Find the "anomaly" by visual inspection.

    c. How did you recognize this anomaly? What is the statistical principle?

    d. Argue against application of a *t* test to verify the anomaly.

    e. From the table, what seems to be the immediate cause of this anomaly?

    f. What might underlie the immediate cause of (e)?

    g. What other features of the data strike your eye?

Rayleigh's values of density of nitrogen

Date	Source	Density
29 Nov. 1893	Nitrous oxide	2.30143
2 Dec. 1893	Nitrous oxide	2.29890
5 Dec. 1893	Nitrous oxide	2.29816
6 Dec. 1893	Nitrous oxide	2.30182
12 Dec. 1893	Air	2.31017
14 Dec. 1893	Air	2.30986
19 Dec. 1893	Air	2.31010
22 Dec. 1893	Air	2.31001
26 Dec. 1893	Nitric oxide	2.29869
28 Dec. 1893	Nitric oxide	2.29940
9 Jan. 1894	Ammon. nitrite	2.29849
13 Jan. 1894	Ammon. nitrite	2.29889
27 Jan. 1894	Air	2.31024
30 Jan. 1894	Air	2.31010
1 Feb. 1894	Air	2.31028

NOTE: Data from "On an anomaly encountered in determinations of the density of nitrogen gas" by Lord Rayleigh (1894), *Proceedings of the Royal Society of London, 55*, 340-344.

1ANS. The two main points of this Exercise are (i) to help develop skills of visual inspection and (ii) to illustrate the common sense of the statistical principle of comparing variability between groups to variability within groups.

a,b. I graphed the points as a function of serial position in time. This showed four distinct groups of data points; the measured density is not constant, which is certainly anomalous. The data may be graphed in other ways, for example, by source of nitrogen.

c. The statistical principle is that variability within each group of points is much less than variability between these groups. Hence the difference between groups must be real.

d. When the data are so clear, a formal significance test is unnecessary. When the data are so clear, reporting a significance test tells your readers that you lack maturity.

e. The immediate cause of the anomaly seems to be whether the nitrogen comes from air, or from one of the three nitrogen compounds.

f. The natural interpretation is that this anomaly comes from some shortcoming in procedure that yields impure nitrogen in one or both classes of samples. As one of numerous possibilities, some oxygen might have remained in the nitrogen derived from air, which would account for the direction of result. As another, some contaminant might remain in the nitrogen prepared from the nitrogen compounds, or in the nitrogen prepared from the air. One of Rayleigh's checks was to use a different method of purifying the nitrogen derived from air for the last three measurements. As can be seen, they give the same result. See also Note below.

g. The first group of four data points shows substantially higher variability than any other group. Perhaps practice made the procedure more uniform and reduced the variability for the later measurements.

NOTE: At that time, air was considered to consist solely of oxygen and nitrogen, which had revealed themselves through chemical reactions and had been satisfactorily established as true elements. In another paper, Rayleigh noted that he had originally used one method of get pure nitrogen by burning out the oxygen, and having a obtained a few consistent results with it, thought his work was complete. But just to be sure, he tried another method and found to his "disgust and impatience" that he got a different result. Thinking he had overlooked some consideration in the chemistry, which was not his field, he published a note in *Nature*, "inviting criticisms from chemists who might be interested in such questions." He got a number of replies, but nothing too useful.

Rayleigh could have dropped the matter, leaving it for someone else to worry, for it seemed just a nuisance complication. Instead, he enlisted the collaboration of a chemist, Sir William Ramsay, with whose aid he obtained the above measurements. With extensive effort, they were able to rule out numerous alternative explanations and conclude that the cause of the anomaly was that the air contained a small amount of a mysterious component, hitherto unsuspected because it was chemically inactive. With additional effort, they provisionally identified this component as a new element, to be called argon. This was a notable feat since argon was chemically inert, and could not be revealed with chemical methods. In fact, Rayleigh received the Nobel prize in physics in 1904, explicitly for the discovery of argon. Ramsay received the Nobel prize in chemistry at the same time.

Rayleigh's discovery was serendipitous in that air was used in both methods in his original work. The second method, however, had initially passed the air through liquid ammonia, which contributed some nitrogen not accompanied by argon. Hence the net density was less. This discrepancy was of course only a fraction of that in the above table. Insightfully, Rayleigh seized on this clue, and followed it up by using nitrogen from nonair sources. See similarly *Confounding Wins Nobel Prize* in Section 8.1.6.

Also of interest, Ramsay used Mendeleyev's 1869 periodic table to predict the existence of at least three more chemically inactive and inert gases, which he went on to verify (neon, krypton, and xenon; and later radon).

2. An experiment with $n = 10$ subjects in each of two groups yields $\overline{Y}_E = 8$ and $\overline{Y}_C = 4$, with $s = 3$.

    a. Show that the 95% confidence interval for the mean difference is 4 ± 2.82.

b. Show that the *t* ratio for the mean difference is 2.98 (see Note 2.2.0a).

c. Compare and contrast the implications of (a) and (b).

2ANS. a. On 18 df, $t^* = 2.10$ for $\alpha = .05$. From Expression 2, the confidence interval for the mean difference has width $\pm \sqrt{2} \times 2.10 \times 3 / \sqrt{10} = \pm 2.82$.

b. $t = 4\sqrt{n} / \sqrt{2} \, s = 2.98$.

c. Both the confidence interval and the *t* test show that the difference between means is statsig; the former because 0 is outside 4 ± 2.82, the latter because 2.98 is greater than 2.10. The confidence interval goes further, however, to put bounds on the likely error of the mean difference.

3. By visual inspection of Figure 2.1, estimate power for $\alpha = .05$ and also for $\alpha = .01$, assuming the given curve for H_0 false applies.

4. Consider the sample {1, 1, 1, 2, 2, 2, 3, 3, 3, 4, 4, 4, 5, 5, 5}.

a. Show by visual inspection that the sample mean is 3.

b. Calculate the sample variance by hand from Equation 4.

c. Show that the 95% confidence interval for the sample mean is $3 \pm .81$.

d. Construct the 99% confidence interval for the sample mean.

e. Explain the difference in width of these two confidence intervals.

f. In your opinion, is the 4% increase in confidence worth the increase in width of the interval?

4ANS. a. Visual inspection shows that the sample is symmetrical, with 3 as the center. Thus, the 1s and 5s average to 3, and similarly the 2s and 4s.

b. $s^2 = [3(1-3)^2 + 3(2-3)^2 + 3(3-3)^2 + 3(4-3)^2 + 3(5-3)^2]/14 = 30/14 = 2.1429$. Accordingly, $s = 1.464$.

c. On 14 df, $t^* = 2.145$ for $\alpha = .05$. Hence the 95% confidence interval for the sample mean is $3 \pm 2.145 \times 1.464 / \sqrt{15} = 3 \pm .81$.

d. On 14 df, $t^* = 2.977$ for $\alpha = .01$. The confidence interval is thus 3 ± 1.13—40% wider.

5. Prove the assertion of Section 2.2.7 that if 95% of the sample means lie within 2 standard deviations of the population mean, then an interval of width ± 2 standard deviations about the sample mean assures you 95% confidence that this interval contains the population mean.

5ANS. Given: 95% of the sample means from a certain population lie within 2 standard deviations of the population mean. Of course, if any sample mean lies within 2 standard deviations of the population mean, then the population mean lies within 2 standard deviations of that sample mean. Hence 95% of the confidence intervals contain the population mean. Accordingly, we may have 95% confidence that our particular interval contains the population mean.

6. What does the text mean by saying in Section 2.3.4 that "Every experiment is predicated on the assumption it has adequate power"?

6ANS. You don't ordinarily do the experiment unless you have a reasonable expectation of getting a statsig result. The surface reason is that nonstatsig results don't get much attention, especially from journal editors. The deeper reason is that this lack of attention is justified. Low power means that the real effect is small relative to the error variability so results will seem inconsistent from one experiment to the next. The "long shot" experiment is no exception. The long shot refers to a long chance of a substantial real effect—with substantial power—not to a small effect with a long chance of being statsig. A possible exception to the text statement might arise with a person engaged in a critical experiment in which a statsig result would infirm his theory.

7. You are TA in an undergraduate class on research methods.

a. Write a paragraph for your students giving an intuitive rationale why larger samples have narrower confidence intervals for the sample mean.

b. Show how this intuition is quantified with a formula in the text.

7ANS. a. A larger sample carries more information. It gives more opportunity for random fluctuations to average out, thereby reducing the likely error of the sample mean.

b. This greater informativeness is quantified in the law of sample size for the standard deviation of the sample mean, Equations 7b,c.

8. You are TA in undergraduate research methods. Write a paragraph giving your students an intuitive explanation about the common sense of "likely error" as discussed in the fourth paragraph of Section 2.1.2.

8ANS. A sample mean will differ from the population mean because it consists of a limited number of elements, chosen by chance from the population. The "likely error" of the sample mean can be viewed as an average of the differences between means of different samples. (This average is quantified as the standard deviation of the sample mean.) Our problem is to estimate this likely error with just a *single sample*.

The key is to look at the variability of the numbers in our single sample. We expect similar variability in other samples. If the variability in our sample is small, therefore, we expect another sample to have a mean close to our mean; if the variability in our sample is large, we expect another sample to have a mean not close to our sample mean.

Fortunately for empirical scientists, this commonsense reasoning can be fulfilled with an exact formula for confidence intervals.

9. In what way does the standard deviation of the sample mean distribution differ from the standard deviation of the population distribution?

9ANS. The standard deviation of the sample mean distribution is smaller than the standard deviation of the population by $1/\sqrt{n}$ (Equation 7b).

10. a. In what *qualitative* ways will the sample mean distribution for samples of size 3 differ from that for samples of size 2?

b. In what way will they be the same?

10ANS. a. The sample mean distribution has smaller variance for samples of size 3 than for samples of size 2. It will also be more normal (unless the population itself is already normal).

b. Both will have the same mean.

11. What relation is there between the distribution of heights of a population of adult women and the corresponding sample mean distribution for:

   a. samples of 1 woman?   b. samples of 2 women?
   c. samples of *n* women?

11ANS. a. For samples of size 1, the sample mean equals the sample element itself. The sample mean distribution thus has the identical shape as the population distribution.

   c. For samples of size *n*, the sample mean distribution:

   i. has a mean equal to the population mean (Equation 4);

   ii. has a variance $1/n$ of the population variance (Equation 5a);

   iii. is more normal (unless the population is already normal) by the central limit theorem.

12. This exercise concerns similarities and differences between a jury trial and a significance test. *Note to instructor*: I suggest assigning only a few parts of this question at a time.

   a. What is the legal analogue of the false alarm parameter, α?

   b. What are the legal analogues of miss and false alarm?

   c. P says the analogue of the null hypothesis is "The defendant is guilty." Q argues for "The defendant is innocent." What do you say?

   d. What is the legal analogue of the stricture, "Do not accept H_0?

   e. What is the legal analogue of the $\alpha-\beta$ tradeoff? How does the legal system handle this tradeoff?

   f. What is the legal analogue of power?

   g. How does power affect the behavior of prosecuting attorneys?

   h. How does power affect the behavior of defense attorneys?

   i. Do you see any logical difference between the decision of a jury and the decision of the journal editor on your thesis you submit for publication?

   j. Jurors may judge whether the defendant deserves to be punished by taking extenuating circumstances into account, not merely whether he or she is guilty of having performed a certain action. In your opinion, how much does this consideration change the foregoing decision analysis?

(As a case with historic interest, Daniel Sickles shot and killed his wife's lover, not under the compulsion of momentary emotion, but with deliberate premeditation. There was no doubt about his action; nor that such action was criminal. Yet he was acquitted by the jury. Perhaps the jury would not have bent the law so far had they foreseen that in his later career as a brash, political general in the Civil War, Sickles would put the Union army in dire peril at the Peach Orchard at Gettysburg on 2 July, 1863. Some *truly wonderful* letters by his wife are quoted in *Sickles, the Incredible*, Swanberg, 1956.)

12ANS.

a. The legal analogue of α is "beyond reasonable doubt."

b. A miss corresponds to a defendant who is guilty but is found "not guilty." A false alarm corresponds to a defendant who is innocent but is found "guilty."

c. Q is correct.

d. The jury decides "Guilty" or "Not Guilty." "Not Guilty" does not mean "Innocent," which corresponds to H_0.

e. The legal analogue of α can be changed by changing the value of "reasonable" in "beyond reasonable doubt." A more stringent "reasonable" will decrease the likelihood of convicting both innocent and guilty. The $\alpha-\beta$ tradeoff is sometimes expressed in the form "Better 10 guilty go free than 1 innocent be convicted." As far as I know, however, the meaning of "reasonable" is left to the jurors.

f. The legal analogue of power is the weight of the evidence against the defendant.

g. Prosecuting attorneys tend to avoid prosecuting cases with low power (weak evidence), partly to avoid waste of time, partly to avoid failures on their records.

h. Defense attorneys naturally dislike cases with high power. One tactic is to seek a settlement or offer a "plea bargain," in which the defendant pleads guilty to some lesser offense to avoid the risk of standing trial on a greater offense.

i. Figure 2.2 applies equally to both the jury and the journal editor. Costs and benefits differ in the two cases, and these will govern the choice of α.

There is one essential difference. The journal editor and reviewers will take prior evidence into account, whereas juries are instructed to presume innocence and prior record of the defendant may be inadmissible.

j. I suggest that essentially the same decision analysis holds, except that the dimension of judgment underlying the decision is not whether the person committed the act but rather a dimension of deservingness of punishment.

13. You study efficacy of prayer by having devout persons pray over corn seedlings, using a *t* test to compare their growth rate with a control.

   a. Your data analysis yields $t(60) = 2.00$. How confident are you *personally* that *prayer* had the observed effect? How confident are you *statistically*?

   b. You do an exact replication of the experiment and get similar results. Now how confident are you personally that prayer had the observed effect?

14. An undergraduate in your class on research methods measures the stocking-foot height of all persons in a certain class and constructs a confidence interval. She finds mean height statsig greater for females than males. What different explanations would you consider possible?

14ANS. 1. Arithmetic error.   2. False alarm.   3. Seventh grade class.

15. What mental model do you think underlies the intuitive feeling that larger populations require larger samples to get same accuracy?

ANSWERS FOR CHAPTER 3

NOTE. Exercises have varied levels of difficulty. Some ask for calculations that have precise answers. Others ask for conceptual interpretation or for personal research judgment and often do not have precise answers; these exercises address issues that are your main concern in actual research. Developing your research judgment is a continuing matter, for which these exercises are stepping stones.

1. Given these three groups of scores:

   group 1 {3, 4, 5}; group 2 {5, 6, 7}; group 3 {6, 7, 8}.

   a. Get s^2 for group 1 using pencil and paper.

   b. By visual inspection, say why the other two groups also have $s^2 = 1$.

   c. What principle underlies the rationale of (b)?

   d. Calculate $F = 7.00$ by hand, using the hand formulas of Section 3.2.5.

(F and t ratios are reported to two decimal places, unless F is larger than, say, 20. Accurate F and t ratios often require *at least* four digit accuracy in interim calculations.)

1ANS.

a. For group 1,
$$s^2 = [(3-4)^2 + (4-4)^2 + (5-4)^2]/(3-1) = 1.$$

b. Visual inspection shows that the *differences* from the group mean for the three successive scores in group 2 are identical to those in group 1; and similarly for group 3. These *differences* determine s^2. Hence $s^2 = 1$ for these groups also.

c. The principle in (b) is that adding a constant to all the scores leaves the variance unchanged.

d. $SS_{between} = 3[4^2 + 6^2 + 7^2] - 9(17/3)^2 = 14.$
Since $MS_{within} = s^2 = 1$, $F = 7.00$ on 2/6 df.

2. Suppose the example of Section 3.2.3 had the scores {1, 3, 5} and {8, 8, 11}.

   a. By visual comparison with the example, show that $MS_{within} = 3.5$.

   b. By comparison with the example, guess roughly at the value of F.

   c. Calculate F by hand and compare it to your guess in (b).

2ANS. a. Visual comparison with the numerical example shows scores are unchanged for group 1, 1 point higher for group 2. Hence MS_{within} is the same as in the example, namely, 3.5.

c. To calculate $SS_{between}$, note that $\bar{Y}_1 = 3$, $\bar{Y}_2 = 9$, and $\bar{Y} = 6$. Hence $SS_{between} = 3[3^2 + 9^2] - 6 \times 6^2 = 54$. Accordingly, $F = 54/3.5 = 15.43$ on 1/4 df.

3. Growing up is as difficult as it is important, and we get no second chance. Research on parenting should thus be a preeminent concern of psychological science. This is far from true, which seems to me a grave criticism of our field.

Some pioneer work has been done, however, to compare training programs for parents of children with developmental disabilities. Two parent training programs are compared in Figure 3.3. Each black bar represents one parent trained with a

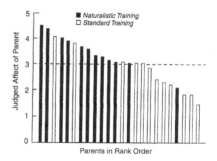

Figure 3.3. Parents of autistic children show more positive affect when training their child using naturalistic procedure (black bars) than procedure based on standard learning theory (white bars). (After Schreibman, Kaneko, & Koegel, 1991.)

naturalistic procedure, which emphasized flexible interaction between parent and child, aiming especially to increase the child's motivation, as by allowing the child to participate in choice of activities. Each white bar represents one parent trained with a structured procedure, following precepts of standard learning theory, which emphasized giving the child well-structured tasks, clear instructions, and trial-by-trial reinforcement with reinforcers chosen to be functional for each child. The main training material was in a manual studied by each parent, who was monitored to a specified criterion of performance with their own child. Naive blind observers watched videotapes of parent–child interaction; they rated parent's affect from negative (0) to positive (5).

   a. By visual inspection of Figure 3.3, do you think the difference between training programs is reliable? Why *exactly* do you think so?

   b. Mean affect is 3.55 and 2.72 for the two groups, with $n = 12$. $\sum Y^2 = 250.681$. Calculate F by hand. From Expression 2 on page 34, show that the 95% confidence interval for the mean difference has width $\pm .59$. Interpret the results.

   c. How much, if anything, does the statistical analysis of (b) add to the visual inspection of (a)?

   d. What do you think of the communication power of Schreibman's graph?

3ANS. b. $SS_A = 12[3.55^2 + 2.72^2] - 24[3.135^2] = 4.134.$
$SS_{error} = 250.681 - 240.011 = 10.670.$ $MS_{error} = .485.$
Thus, $F_A = 8.52$, which is greater than $F^*(1, 22) = 4.30$. The mean difference is $3.55 - 2.72 = .83$ so the 95% confidence interval is $.83 \pm \sqrt{2} \times \sqrt{4.30}\sqrt{.485/12} = .83 \pm .59.$

This result is important because parents who experience positive affect are more likely to continue the training procedure in their everyday interaction in the home. The better showing of the naturalistic procedure is interesting because it should be more easily incorporated within the everyday family life. Some evidence, it may be added, suggests that the children learn more under the naturalistic procedure.

c. If you have developed your skills of visual inspection, this striking graph should suffice. Even so, Anova adds definiteness. And it avoids forcing the reader to worry about the matter. In addition, of course, this statistical analysis would be indispensable for calculating power for further work (Section 4.3).

d. I consider Schreibman's graph a model of communication, which we should all seek to emulate. The entire distribution of both samples is visible in a readily comparative mode, which shows, among other things, that there were no extreme subjects.

4. Supplementary analyses are sometimes desirable on a published article. Suppose a journal article reports means of 3 for a control group, and 7, 9, 11 for three experimental groups, with $n = 8$. The reported $F(3, 28)$ is 8.00, so the author concludes the data show real differences. You wish to test differences among the three experimental groups. Use hand calculation with the formulas of Section 3.2.5.

a. Show that $MS_{error} = 11.667$.

b. Do Anova for the three experimental groups.

c. Discuss the relation between your analysis and that in the article.

4ANS.
a. $SS_A = 8[3^2 + 7^2 + 9^2 + 11^2] - 32 \times (30/4)^2 = 280$. $MS_A = 280/3 = 93.333$. $MS_{within} = MS_A/F_A = 11.667$. $SS_A = 8[7^2 + 9^2 + 11^2] - 24 \times 9^2 = 2008 - 1944 = 64$. $MS_A = 64/2 = 32$; $F_A = 32/11.667 = 2.74$, which is less than $F^*(2, 28) = 3.34$.

c. Your analysis shows the data are inadequate evidence for real differences among the three experimental conditions. This does not mean you should conclude there is no difference; the evidence gives some support but not enough for reasonable confidence. If the main concern is with differences among the three experimental conditions, your analysis is the proper one. On the other hand, if the main concern is with the differences between experimental and control conditions, the analysis in the article is appropriate.

5. By visual inspection, guess whether F for each of the following three data sets would be larger, smaller, or equal to the F for Exercise 1.

a. {103, 104, 105};   {105, 106, 107};   {106, 107, 108}.

b. {4, 5, 6};   {6, 7, 8};   {7, 8, 9}.

c. {-8, -7, -6};   {-6, -5, -4};   {-5, -4, -3}.

d. What principle underlies these examples? Why is this a good principle?

5ANS. a,b,c. In all three cases, the three sets of scores differ from those in Exercise 1 by the same additive constant. Hence SS_{within} is unchanged because the *differences* among the numbers within each group are unchanged. Also, $SS_{between}$ is unchanged because the *differences* between the group means are unchanged. Hence also F is unchanged.

d. The principle is that adding a constant to all scores leaves the *differences* among the numbers unchanged. Hence MS_{within} remains unchanged, and hence also F. This is a good principle because otherwise you would get different results depending on whether you measured temperature, for example, in Celsius or Fahrenheit.

6. In Exercise 1, suppose you change one score by one point.

a. What score would you change to maximize F?

b. What score do you change to minimize F?

6ANS. a. To maximize F by changing one score one point, change 5 to 4 in group 1 (or 6 to 7 in group 3). No other one-point change can produce a greater change among the group means. At the same time, either of these changes decreases the error variance and no other one-point change can do better in this regard. For example, changing 3 to 2 in group 1 will increase the mean differences equally but will also increase the error.

b. To minimize F, change 5 to 6 in group 1 (6 to 5 in group 3).

7. P says, "I'm not interested to test whether the effect is statsig; I just want to know how big it is." What circumstances—if any—do you think would justify P's attitude?

7ANS. Previous knowledge may indicate that there is definitely a real effect, for example, the relation between age and vocabulary size in children. A significance test to demonstrate a real effect would be obtuse here.

Alternatively, the effect may be obvious when P graphs it. In this and the preceding situation, of course, P will no doubt include error bars in the graph. I am sure, however, that he will resist pressure to intrude a formal significance test when it is superfluous. Such situations are not infrequent in perception (e.g., line–box illusion in Figure 8.1, page 226, and iguana perception in Exercise 21 of Chapter 6).

In some cases, one of several given treatments of equal cost must be adopted. Lacking other information, the clear choice is the treatment showing the largest effect, regardless of statistical significance.

8. You are TA in the advanced undergraduate statistics course. The students are assigned to read the subsection on *Robustness of* α in Section 3.3.6. Three students come to your office confused about the meaning of "Ideally, this proportion will equal α" in the third paragraph. What do you tell them?

8ANS. Suppose you guys did an experiment with data distributed according to the bimodal, lumpy distribution shown in Figure 3.2. You apply Anova and submit your paper for publication. One reviewer recommends against publication, arguing that Anova assumes normal distributions and yours are far from normal. Hence, says the reviewer, you do not really know what your operative α is; it might be .12 or higher.

If Anova is robust against nonnormality, however, then your operative α will be near .05. Ideally, it will be .05 exactly. What Table 3.2 shows is that the ideal is nearly true; the proportion of statsig tests in the 10,000 simulations for this bimodal, lumpy distribution was close to .05.

9. Those same three students return the next week, now confused about *Robustness of Power*. What, they ask, does it mean

to say that "Relative to the normal distribution, power was roughly .03 less."

9ANS. Even though Anova keeps α near .05 for the bimodal distribution of Figure 3.2, it might suffer from loss of power to detect real effects. That is, β might be affected. Sawilowsky and Blair calculated power for four sizes of real effects for samples drawn from this bimodal distribution. But to assess how much power is lost because of the nonnormality, some comparison standard is needed.

They obtained this comparison standard by calculating power for a normal distribution with the same size real effects and the same error variance. This allows estimation of how much power the nonnormality costs. For the eight distributions they considered, power loss was small, roughly .03 for the bimodal, lumpy distribution.

10. The Rev. Franklin Loehr (1959) has made a sincere, dedicated effort to place religion on a scientific foundation. Prayer is considered efficacious by many, but experimental evidence is scanty. Loehr saw prayer as an ideal field for experimental analysis. He used seeds of fast-growing plants (e.g., corn) and had devout persons pray over them as they germinated and sprouted. The response measure was the distance from the soil level to the tip of the highest leaf. Seeds were assigned at random to the prayer and no-prayer conditions. Care was taken to equalize moisture, light, and other environmental factors.

The results were claimed to show a reliable effect; the experimental plants showed more growth than the controls.

a. Why do you think fast-growing plants were selected?

b. One skeptic considered Loehr's design–procedure sound and the data analysis correct. Still, he scoffed that the statsig result must be a false alarm. Is this a reasonable reaction to a bona fide experimental analysis?

c. Rev. Loehr replicated the experiment and found again that the prayed-over plants did statsig better. In light of this replication, is there any reason the skeptic of (b) should not be at least moderately convinced of the positive power of prayer on plants?

d. Suppose the result was not near statsig, even with a large N. Of itself, without regard to background knowledge, how strongly would this negative outcome argue against the positive power of prayer on plants?

Note: Loehr's (1959) popular book gives little technical detail. Various reasonable precautions were said to have been taken and in at least one case the person who measured the plants was blind to their treatment condition. One curious claim was that negative prayer had greater effect than positive prayer. Technical articles are said to be published in *Religious Research*, but I have not read them. Galton (1872) found that people who were heavily prayed for (e.g., English monarchs) did not live longer, but the ingenuity of his inquiry does not mitigate the confounding. The apparent lack of experimental analysis of this issue seems remarkable.

10ANS. a. Fast-growing plants were (presumably) selected to reduce the time cost and to amplify the opportunity for prayer to act.

b. Treating a statsig result as a false alarm is a reasonable action

for those who strongly doubt that prayer will really affect plant growth.

c. If the replication is successful, the skeptics should be moderately convinced that the result is not a false alarm. This does not mean they conclude that prayer is efficacious. Instead, they would look for confounds that accompanied the prayer that could account for the observed difference. The most obvious confound would be whether the person who measured the plants was blind about which had been prayed over, which not. Ample evidence shows that nonblind measurements cannot be trusted (e.g., Section 8.1).

d. A negative outcome would have small evidence value. Any number of experimental details could account for a negative outcome. Perhaps slow-growing plants would show an effect; perhaps prayer would be effective at a later stage of plant life; perhaps those who did the praying were ineffectual pray-ers; perhaps a different kind of prayer is needed; perhaps the control seeds were too close to the experimental seeds and benefited equally from the prayer.

11. Suppose three group means are 3, 5, and 7, with a total of 30 subjects. The error mean square is 9.

a. Use the hand formulas of Section 3.2.5 to show that F is 4.44 if there are 10 subjects in each group, and 2.99 if the subjects are divided 3, 10, and 17, for the three groups in the given order.

b. What moral does this exercise suggest?

11ANS. a. For $n = 10$,
$SS_A = 10[3^2 + 5^2 + 7^2] - 30 \times 5^2 = 830 - 750 = 80$.
Hence $MS_A = 40, F_A = 40/9 = 4.44$.

With ns of 3, 10, and 17,
$SS_A = 3 \times 3^2 + 10 \times 5^2 + 17 \times 7^2 - 30 \times 5.933^2 = 53.88$.
Hence $MS_A = 26.93, F_A = 2.99$.

b. The moral is that equal n yields more power (at least with equal variance).

12. After presenting both a hand calculation and a Minitab printout for a t test, Howell (1992, p. 171) points up a presumed advantage of Minitab, saying "Thus, whereas we concluded that the probability of a Type I error [= false alarm] was *less than* .05 [with the hand calculation], Minitab reveals that the actual probability is .0020." [Here .0020 is the p value on the Minitab printout.]

Explain Howell's misconception (consider H_0 false and H_0 true separately).

12ANS. Howell identifies the p value with the probability of a false alarm (Type I error). This is a misconception. The p value is the probability that a t as great or greater would be obtained **if H_0 was true.** But if H_0 is not true, a false alarm cannot occur. And if H_0 is true, then a false alarm will occur with probability .05, not .0020.

13. Consider the error variability, MS_{within}. Other things equal, increasing sample size will: (a) increase error variability; (b) decrease error variability; (c) neither of these two; it depends. (d) none of these.

13ANS. Increasing sample size leaves MS_{within} unchanged on average. Note that MS$_{within}$ is an estimate of the *population* variability, σ_e^2, which is totally independent of sample size. More concretely, MS$_{within}$ is usually composed mainly of individual differences and these are surely not decreased by adding more individuals. Thus, (d) is correct.

14. You are TA in the undergraduate course on research methods, which is currently covering variance and confidence interval. One of your students comes to you saying, "My variance is negative, and I can't figure out what that means." How do you answer?

14ANS. How you answer depends on what kind of teacher you are trying to be. A quick answer is "You've made an error. Look at the definition of variance in Equation 1b: The variance is a sum of squares, so it can't be negative. Try again."

More effective teaching is to reinforce the student for recognizing that negative variance is suspicious. Then go over the student's calculations with the student to locate the error and try to illuminate its source. A likely source is just carelessness, which should be brought out gently but clearly. Then you can say that the variance can't be negative.

15. Every time you do an experiment, you are taking a random sample (at least ideally). Sooner or later, you are going to be hit with a "bad" random sample.

　　a. In terms of confidence interval, in what way(s) can a sample be "bad?"

　　b. What proportion of samples from a normal distribution will be "bad."

　　c. Why not just throw out the "bad" random samples?

15ANS. a. The most obvious way a sample can be "bad" is that its mean is far from the population mean. More specifically, a "bad" random sample is one for which the confidence interval around the sample mean does not contain the population mean. A sample can also be "bad" by giving a too-small (too-large) variance and hence a too-narrow (too-wide) confidence interval.

b. Exactly α of the samples from a normal distribution will be "bad."

c. We can't throw out the "bad" random sample because we don't generally know which ones they are (even though α are guaranteed to be "bad.")

　　Bad random samples are sometimes detectable, however, by comparing the sample results with previous knowledge. In particular, an effect contrary to previous knowledge should be considered a possible "bad" random sample.

16. Write a one-page essay on the concept of "placebo," using the index entries in the Subject Index.

Exercises on Random Assignment

r0. Some people claim that different groups of subjects should be carefully matched on relevant characteristics before beginning an experiment: "Leaving this critical matter to chance is utter madness; the proof of this madness is that the randomizers admit that random assignment *guarantees* a proportion α of false alarms (given H_0 true)." This claim sounds pretty reasonable, don't you think? What can you add to support this claim?

r0ANS. See Section 14.2.

r1. a. Construct a subject assignment sheet with 16 subjects in each of three conditions, using unblocked randomization. For purposes of this problem begin at the first entry in Table A1 of random numbers and proceed lexicographically. Let digits 1 to 3 denote the three experimental conditions; ignore the other digits. Record your stopping location by row and column.

　　Include space on your subject assignment sheet for all information that is to be collected in the experimental session (see text).

　　b. What seemingly nonrandom features can you detect in your sequence?

r1ANS. a. The subject assignment sheet should include a line or two for each subject, with ruled space for experimental condition, subject name, date and experimenter, comments, and labeled with the name of the experiment.

b. Random sequences are typically more irregular than one expects. Most people are surprised how often the same treatment is bunched together, with longish gaps between bunches. With nonblocked randomization, moreover, one or two conditions will probably be filled early, so the last group of trials may contain a seeming overplus of one or two conditions.

r2. Construct a subject assignment sheet with 10 subjects in each of 4 conditions, randomized in blocks of 8, using one permutation of 16 numbers in Table A2 for each block. In practice, the first permutation would be chosen haphazardly and subsequent permutations lexicographically. For this problem, however, begin with the first listed permutation and proceed lexicographically. For convenience, assign numbers 1 and 2 to condition A_1, numbers 3 and 4 to condition A_2, and so on, in each permutation. Ignore the remaining numbers.

r2ANS. If the first permutation was

　　2　16　8　14　9　3　10　6　7　1　5　11　4　13　15　12,

the first eight subjects would be run in the order

　　A_1　A_4　A_2　A_3　A_4　A_1　A_3　A_2.

r3. To her horror, Q realized at the end of her experiment that all seven subjects in her A_1 condition had turned out to be male, all seven in her A_2 condition female.

　　a. Q had used a table of random numbers and made up a subject assignment before scheduling any subjects. Was her randomization procedure at fault?

　　b. What should Q do now?

r3ANS. a. No and Yes on faulty randomization.
No. The 7–7 split is unlikely, but not too unlikely.
Yes, her randomization was at fault for not blocking on gender in constructing her subject assignment sheet. With blocking, number of males and females could be be split 3:4 in both conditions. Even *n*, say 4 of each gender, would have advantages.

b. If Q thinks that the gender confounding is not important, she can write up the results, explicitly mentioning the confounding, and submit it for publication. The reviewers and editor can decide whether to accept her judgment. This action would seem appropriate if, for example, the subjects were from an unusual population difficult to replicate.

On the whole, however, replication seems desirable. Even if the gender confounding seems innocuous, it is better workmanship to leave no blotches on the work. By using block randomization in the replication, Q can test the main effect of gender as well as its interaction residual with conditions using factorial design as shown in Chapter 5.

r4. Write down how to do a block randomization for Q in Exercise r3.

r4ANS. Q could block on gender in various ways. It is preferable to equalize numbers of females in both conditions, say four, leaving three males in each condition. (This avoids complications from unequal n, Section 15.4.) Use one permutation of 16 numbers from Table A2. Let 1 to 4 denote females in condition A_1, 5 to 8 denote females in condition A_2. Let 9 to 11 denote males in condition A_1, 12 to 14 denote males in condition A_2. Within gender, the first subject is assigned the condition specified by the first number in the permutation, and so on.

r5. You plan to run three groups of undergraduates from the subject pool, for which you post a sign-up sheet. You have no control over which subject appears at each specified time. How can you randomly assign subjects to conditions?

r5ANS. Randomization can be accomplished by using a previously prepared random assignment sheet, as described in the text. What order the subjects arrive for the experiment is then irrelevant.

r6. Subject randomization might be done online manually, for example, by throwing a die or tossing a coin as each subject enters the laboratory. What objections might there be to this (assume block randomization is used)?

r6ANS. Possible errors in subject assignment are more likely with online randomization. Also, online procedure leaves open temptation to change assignment depending on the investigator's hypothesis and the nature of the subject. This possibility is the more to be avoided the more it may be tempting. More than one helpful research assistant has provided results that could not be replicated.

r7. Under *Bad Randomizations* in Section A.1, can rejecting the given "bad randomization" bias the treatment means?

r7ANS. In this instance, rejecting the bad randomization cannot bias the treatment means because it is done before collecting any data and because the block randomization is unbiased. If there is some temporal trend, the given order would confound this trend with the treatment means and so is better avoided.

r8. McGuigan (1993, pp. 69-70) cites the case of a "knowledgeable graduate student" who wished to compare speed of rat maze running for an experimental and a control group. It came out that he reached into the cage of rats and assigned the first rats that came to hand to the experimental group, the others to the control.

a. What bias seems not unlikely with this subject assignment procedure?

b. How serious do you consider this possible bias?

c. What would be the simplest way to get randomized assignment?

d. What are two morals of this example?

r8ANS. a. Rats the investigator can readily grab seem likely to differ in activity level, fearfulness, and so on. Such characteristics may well influence running speed.

c. Before selecting the first rat, make out a randomization sheet that lists which treatment condition is to be given each successive rat as it is captured. Then the order in which rats are captured is immaterial, for it is randomized over conditions. Randomization within temporal blocks may of course be generally desirable.

d. One moral is always use a random assignment sheet. Another moral is to instruct your research assistants carefully and explicitly (see Note 8.1.5c).

r9. Clinical research often has difficulties in getting randomized experimental and control groups. The *waiting list control* procedure (Kazdin, 1992, p. 124*ff*; Fowles & Knutson, 1995; see also Zhu, 1999) begins by asking those who apply for therapy whether they would participate even if treatment were delayed. Those who agree are randomly assigned to an experimental group, in which therapy begins at once, and a control group, in which therapy is delayed some specified time. An assessment of the controls just before their treatment is begun is compared to an assessment of the experimental group taken at the same time.

a. What could go wrong with this waiting list control procedure?

b. What alternative design might be workable?

r9ANS. Some who agree to defer therapy will change their minds during a delay. One alternative is to compare two modes of therapy using random assignment. No current mode of therapy works so well that no improvements are worth testing out.

ANSWERS FOR CHAPTER 4

1. In what way does the significance test of the difference between two means deal with the following questions? (After Freedman, Pisani, & Purves, 1998.)

    a. Was the experiment well designed?

    b. What is the substantive meaning of the result?

    c. Is the difference important?

    d. What is the size of the difference?

    e. Is the difference due to chance?

1ANS. a,b,c. The significance test says nothing about these questions.

d. Although the significance test does compare the observed difference to the chance difference, the chance difference depends also on sample size, whereas the substantive effect size does not because it is a property of the population.

e. The significance test assesses whether chance alone would be likely to produce this effect. But this differs subtly from saying that a significant result implies that some other cause is likely.

2. a. In what ways could a very large effect fail to be statsig?

    b. In what ways could a very small effect be statsig?

    c. In what ways could an unimportant effect be highly statsig?

2ANS. a. A very large effect could fail to be statsig because the error term is very large or because the sample is very small.

b. A very small effect could be statsig because the sample was very large; or by chance, that is, essentially a false alarm.

c. An unimportant effect could be highly statsig because N was very large, as often happens with tiny effects in large-scale questionnaire studies, or because it is a large effect but obvious/trivial. A rather different reason is that it is uninterpretable from confounding, as with the body–mind correlations cited at the end of Section 1.2.3.

3. Exercise 1 of Chapter 3 considers three groups of scores: group 1 {3, 4, 5}; group 2 {5, 6, 7}; group 3 {6, 7, 8}.

    a. Get the error bar for group 2 using information from the given exercise and assuming an overall Anova. Does this error bar indicate a statsig difference between groups 2 and 3?

    b. Graph the data including error bars with each mean.

    c. Get 95% confidence interval for a single mean.

    d. Get the 95% confidence interval for the difference between every pair of means. What inferences may be drawn from these confidence intervals?

    e. The foregoing data were obtained in a physiological study of emotional arousal by Q, who gave higher levels of a stressor to the higher-numbered groups. Q argued that the difference between groups 2 and 3 was real. What was her argument?

3ANS. a. The error variance is 1 for each group (Exercise 3.1). With $n = 3$, the error half-bar has width $\sqrt{1/3} = .577$. The mean

difference between groups 2 and 3 is 1; this falls well short of the rough rule of three half-bars, so it is not statsig.

c. Since the error variance is 1 for each group, an overall Anova would yield $MS_{error} = 1$. For 95% confidence intervals, therefore, we use F^* on 1/6 df, which equals 5.99. Equally well, we could use $t^*(6) = \sqrt{5.99} = 2.45$. The $\pm$ term in Expression 1b for the 95% confidence interval is thus $\sqrt{5.99} \times .577$, which equals 1.41. Thus, the confidence interval for the mean for any group has width ± 1.41.

d. The width of the 95% confidence interval for the difference between two means is $\sqrt{2}$ times that for a single mean. From (c), therefore, the confidence interval for the difference between two means is just under ± 2.00. Hence the difference between groups 1 and 3 is statsig, as is (just barely) the difference between groups 1 and 2. The difference between groups 2 and 3 is not statsig, as already indicated in (a).

e. Drug concentration is a metric variable, so we may reasonably expect the response to be a continuous function of the concentration (barring a sharp threshold). With this in mind, Q looked at the overall trend. Since the statsig difference between groups 1 and 3 implies a real effect, she considered it reasonable to believe the difference between groups 2 and 3 was also real, even though smaller. Q's argument incorporates substantive considerations into the statistical analysis of the data, a model of scientific inference.

4. Anova between an experimental and a control group yields F a trifle greater than F^*

    a. Where does 0 lie with respect to the confidence interval about the difference between the two means?

    b. Consider the confidence interval for the mean of the experimental group taken alone. Where does the control mean lie with respect to this interval?

4ANS. a. 0 lies a trifle outside the confidence interval for the mean difference.

b. By (a), the control mean is away from the experimental mean by a trifle over $\sqrt{2}$ times the confidence interval for a *single* mean. Hence the control mean lies about 40% outside the confidence interval around the experimental mean alone, as may readily be verified with a numerical example.

5. What are the differences between a real effect and a statistically significant effect? Be precise.

5ANS. A real effect refers to a difference in *population* means. A statsig effect refers to a difference in *sample* means. The concept of statistical significance does not apply to real effects; the true means are different or they are not. The concept of statistical significance applies to sample means, embodying the fact that they are intervals of uncertainty, not definite points.

6. An article claims a statsig difference between two means on the ground that ''$\bar{Y}_1$ was statsig greater than 0, whereas $\bar{Y}_2$ was actually negative.''

a. Use confidence intervals to show the error in this reasoning.

b. Conceptually, what confusion is embodied in the given claim?

6ANS. a. That $\bar{Y}_1$ is statsig greater than 0 implies that the confidence interval around $\bar{Y}_1$ excludes 0. But what is at issue is the difference, $\bar{Y}_1 - \bar{Y}_2$. This confidence interval is $\sqrt{2}$ times wider so it could well include 0, even if $\bar{Y}_2$ is somewhat negative.

b. The conceptual confusion is between the sample mean $\bar{Y}_2$ and the true mean μ_2. $\bar{Y}_2$ being less than 0 is implicitly taken to mean that μ_2 is less than 0, and hence outside the confidence interval around $\bar{Y}_1$.

7. Higher power implies: 1. Greater likelihood of a statsig result. 2. More reliable means. 3. Narrower confidence intervals. 4. Greater likelihood that H_0 is false.

a. Are any of these four answers correct?

b. Which do you consider the best answer?

7ANS a. The first three answers are correct (strictly speaking, (1) assumes that H_0 is false). The last answer is not correct, for power refers to the likelihood of getting a statsig result, *given H_0* false. The "likelihood that H_0 is false" is entirely different from the likelihood of getting a statsig result.

b. Both (2) and (3) are best answers, but (3) seems more informative to me. Narrower confidence interval not only implies more reliable means, but quantifies this reliability.

8. For the experiment you are presently planning, what experimental procedures can you use to decrease error variability?

9. Every sample contains some truth about the population mean as well as some error. How does the confidence interval quantify truth and error conjointly?

9ANS. The confidence interval quantifies truth and error conjointly by condensing the sample information into an interval that specifies likely location of the population mean. The width of this interval represents the error variability—error is part of truth.

10. Uncle Norman's **Golden Marble Test**.

"In my office," says Uncle Norman, "I have an urn with 19 *lead* marbles, one of **gold**. For a modest fee, I will test your null hypothesis by drawing one marble at random." The decision rule is:

> **Gold** marble: Reject H_0.
>
> *Lead* marble: Do not reject H_0.

"The beauty of my test," exclaims Uncle Norman, "is that it maintains α at .05—regardless of the shape of your distribution! No assumption of normality or equal variance is needed!"

a. Does Uncle Norman's test really maintain α at .05 regardless?

b. What else is wonderful about Uncle Norman's Golden Marble Test!

c. In terms of the 2×2 decision table of Figure 2.2, what is

statistically inadequate about Uncle Norman's test?

d. Explain the fatal statistical flaw in Uncle Norman's Golden Marble Test in terms of the two curves of Figure 4.1.

e. What is the moral of this exercise?

10ANS. a. Yes. Uncle Norman's golden marble test maintains the α value at exactly 05—no statistical assumptions are needed beyond random sample. One in 20 random draws will yield a false alarm.

b. All the labor of doing the experiment is completely avoided !

c. In terms of the 2×2 decision table, the statistical flaw in the golden marble test is that $\beta = .95$—regardless of the empirical facts. Power = .05, in other words, no matter what the size of the real effect.

An analysis that ignores the data obviously has some fatal flaw, but the question asks for an answer in statistical terms. This question aims to increase understanding of two basic properties of all tests, namely, α and β.

d. Uncle Norman's Golden Marble Test is analogous to the curve "H_0 true" in Figure 2.1. Its fatal flaw is that the very same curve applies when H_0 is false.

e. This exercise emphasizes the need to consider both α and β in evaluating any statistical test. Although standard Anova is robust, other techniques may have more power when the equinormality assumption is not satisfied, (e.g., trimming in Section 12.1).

11. *A statsig result may not be too reliable.* In an experiment with two groups of 32 subjects, $\bar{Y}_1 = 3.02$, $\bar{Y}_2 = 1.00$, $MS_{error} = 16$.

a. Show that 0 is just outside the 95% confidence interval, a statsig result.

b. Show that power of an exact replication is near .50, not even close to .95.

c. Perhaps the result of (b) is peculiar to this numerical example. Manufacture another yourself, with smaller n and different error variance. Set $\bar{Y}_1 = 1$, and choose $\bar{Y}_2$ so that $t = t^*$. Then calculate power of exact replication.

11ANS. a. $2.02 \pm \sqrt{2} \times 2.00 \times \sqrt{16/32} = 2.02 \pm 2.00$.

b. $\sigma_A = 1.01$; $\sigma_\varepsilon = \sqrt{16} = 4$; f = .2525; $\phi = .2525 \sqrt{32} = 1.43$. Power is about .53.

12. In the numerical example of power for minimum effect size:

a. Show that $\alpha_1 = -3$, $\alpha_2 = 3$.   b. Verify that $\phi = \sqrt{n}/4$.

12ANS. a. Given $\mu_1 = 60$, $\mu_2 = 66$, we get $\bar{\mu} = 63$. Hence $\alpha_1 = -3$, $\alpha_2 = 3$.

b. $\sigma_A = \sqrt{3^2 + (-3)^2/2} = 3$; $f = \sigma_A/\sigma_\varepsilon = 3/12 = 1/4$; $\phi = \sqrt{n}/4$.

13. You are planning an experiment and wonder how many subjects to use. Your guesses about the true means of your four experimental conditions are 4, 6, 9, and 15. Your guess about the error variance is 100.

a. Calculate power for $n = 9$ and $n = 16$ at $\alpha = .05$.

b. What n do you need to get power of .90?

c. Repeat (a) for $\alpha = .01$.

13ANS. a. Power = .47 and .76 for $n = 9$ and 16, respectively.

b. Power of .90 requires $n = 21$ or 22.

c. For $\alpha = .01$: Power = .24 and .53 for $n = 9$ and 16.

NOTE. Power calculations require visual interpretation of closely spaced lines in the power charts, which is not easy. My answers are only approximate, but accuracy is not critical since the power calculations yield only guesstimates for a proposed experiment, not used as a published result.

14. The lead article in *Sex Roles* (1992, Vol. 37) concluded about a hypothesized result: ''These differences were not statistically significant because the sample size was too small.'' Comment.

14ANS. The statement that the effect was not statsig because the sample size was too small presumes that the difference is real. For if the true difference is 0, even a huge sample will not make it statsig. One can understand an investigator who *feels* this way. However, to present this feeling as evidence is confused, to say the least. Replication is needed before publication. Sex is too wonderful to be left to such investigators.

15. A journal article reports a statsig Anova for four groups of 12 subjects, with means of 1, 5, 7, and 8, with $F = 3.86 > F^* = 2.84$.

a. Find the error half-bar for a single mean.

b. Find the 95% confidence interval for a difference between two means.

c. Which of the six possible two-mean comparisons would have 0 outside of the confidence interval for their difference?

d. Find power of an exact replication, taking the given sample data as population values.

e. What qualification on the results of this power calculation are indicated by the outcome of (c)?

f. Suppose A was a metric variable, with increasing levels for the four groups in the order listed. What could be added to the conclusion of (c)?

15ANS. a. The given data are used to find $MS_{error} = 29.79$; the error half-bar is 1.58.

b. The 95% confidence interval for a mean difference is ± 4.50.

c. 1 is statsig less than 7 and 8; no other two-mean comparisons are statsig.

d. Show that $\sigma_A = 2.6810$, $\sigma_\varepsilon = 5.458$, $f = .4912$, $\phi = 1.70$. Power is thus about .78.

e. This power applies only to the overall F, not to the pairwise comparisons. But pairwise comparisons are presumably most important, yet only two are statsig. This power analysis indicates that including all four conditions in a replication may be inadvisable.

f. If A was a metric variable, the appropriate analysis would presumably be a linear trend test. Overall F would not be done.

16. (After Tversky & Kahneman, 1971.) One of your students has performed a difficult, time consuming surgical experiment with 20 rats, with numerous physiological measures. Although her results are generally inconclusive, one before–after com-

parison yields a $t = 2.70$, which is highly significant. This result is surprising and could be of major significance.

Considering the large number of statistical tests, you recommend replication. Your student replicates with another 20 rats. The result is in the same direction, although the t of 1.21 falls short of the criterial value of 2.09. Discuss the appropriateness of each of the following courses of action.

a. Your student should combine the data of the two experiments and publish the result as showing acceptable standards of evidence.

b. She should report the two experiments separately in the same paper, presenting the second as a replication of the first, and treating the combined results as showing acceptable standards of evidence.

c. She should run another group of (how many?) animals.

d. She should try to explain the difference between the two experiments.

e. What alternative course of action would you suggest?

16ANS. A key point is that a large number of comparisons were made in the initial experiment. Hence any statsig result comes under suspicion of being merely due to chance.

a. Pooling the results without making the pooling explicit would be unethical. In particular, it may be taking advantage of a false alarm in the first experiment.

b. Reporting the two experiments separately in the same paper—while making clear the numerous tests made in the first is not unethical. The data have some evidence value that may justify publication, a decision for the reviewers and editor. It would hardly be correct, of course, to treat the results as showing normal standards of evidence. The first t is under suspicion; the second t could be obtained by chance about 1/4 of the time.

c. At a rough guesstimate, more than 50 animals would be needed to get adequate power for a direct replication. The average of the two cited t ratios is 1.955, which is marginal and so suggests power near .50 (Exercise 11). This is likely an overestimate since the first t is not unlikely biased upward by chance fluctuation. Considering that the work is difficult and time-consuming, direct replication seems a bad bet.

d. *There is no difference to explain.* The first result is only suggestive because of the α escalation. The second result was not close to statsig.

NOTE: Tversky and Kahneman included a very similar question on a questionnaire handed out to one group of psychologists at a meeting of the Mathematical Psychology Group and to another group at a convention of the American Psychological Association. Frequencies of choices for (a), (b), (c), and (d) (worded a little differently) were 0, 26, 21, and 30. The most popular choice of (d), as they say, is indefensible.

This confusion in professional researchers is caused in part, I believe, by the prevailing tradition in statistics teaching. More empirics is needed, less statistics.

17. Intuitively, what will happen to the ''H_0 true'' curve in Figure 4.1 if:

a. the treatment effect is increased?

b. error variance is decreased?

c. sample size is increased?

17ANS. a,b. The "H_0 true" curve depends only on the df. It is unchanged by changes in treatment effect or changes in error variance.

c. Increasing sample size increases error df and causes F^* to decrease. Since α is fixed, the right tail of the "H_0 true" curve must shift leftward.

18. Intuitively, what will happen to the "H_0 false" curve in Figure 4.1 if

a. the treatment effect is increased?

b. error variance is decreased?

c. sample size is increased?

18ANS. Increasing the treatment effect or decreasing the error variability must both increase the power index ϕ of Equation 5. Hence both must increase power—the area under the "H_0 false" curve to the right of F^*. Hence both must cause the "H_0 false" curve to shift rightward. Rightward shift will also come from increasing sample size.

19. You are TA in an honors course on research method. One of your student teams finds in their course experiment that the confidence interval for $\bar{Y}_1 - \bar{Y}_2$ excludes 0. "Therefore," they write, "we can be 95% confident that an exact replication of this experiment would yield a confidence interval that also excluded 0."

a. How does Section 4.1.6 disagree with them?

b. How do power considerations relate to this?

19ANS. a. Apply the last two paragraphs of Section 4.1.6, replacing "sample mean" by "sample mean difference."

b. Implicitly, your students are referring to power, that is, the probability of a statsig result—**given** that there is a real effect. But significance statements are conditional on the null hypothesis being true—**no** real effect.

Furthermore, the probability that an exact replication would yield a statsig result may be only .50 (Exercise 11).

20. P and Q conduct parallel experiments, both of which involve the same real effect, and both of which have power of .7.

a. What is the probability their significance tests will be "inconsistent."

b. Why is this not as bad as it sounds?

20ANS. a. P and Q will have "inconsistent" results when one is statsig, the other is not. This probability equals $.7 \times .3 + .3 \times .7 = .42$.

b. The nonstatsig result may still be close to statsig. This outcome has some evidence value, even though not enough to claim a result beyond reasonable doubt. The nonstatsig result could thus add solidity to the statsig result.

If the two results come out in opposite directions, this would be cause for concern even though not gilt-edged inconsistency.

21. Statistics has many uses in industry. Here we consider a simple example from *Mathematical Methods in Reliability Engineering* (Roberts, 1964), in which you must purchase a certain part from one of two vendors. The parts fail after a time, so you wish to select the vendor who manufactures the most reli-

able parts. In your tests, you find that 9 parts from vendor A have a mean life of 42 hours with a standard deviation of 7.48 hours; 4 parts from vendor B have a mean life of 50 hours and a standard deviation of 6.48 hours. Assume price and other considerations are equal. (Thanks to Jaynes, 1976.)

a. In principle, what should you do?

b. Roberts says you should use a t test, and adds:

If, at this juncture, you are tempted to say, without further ado, "I shall choose B. His product exhibits a longer mean life with somewhat less scatter in the data," I can only suggest that you turn back to the beginning of this chapter and start over; or perhaps statistics is not for you!

What do you say to Roberts?

21ANS. a. In principle, you should choose the product with the longer mean life. You must choose one or the other; the given means are the best evidence you have, and on them you must rely. A significance test is utterly pointless for this decision.

b. I say to Roberts: Stop taking "shallow draughts" and "drink deep" at the Pierian spring (see quotation from Alexander Pope at beginning of Chapter 7).

NOTE. Although Roberts' quote is a parody of significance testing, Jaynes cites it as a criticism of Fisher–Neyman–Pearson statistics; and as an argument for the Bayesian school (Chapter 19). Much criticism of statistics rests on similar tactics of citing misuse of some technique as invalidity of the technique itself, not as miseducated users (Note 19.2a).

NOTE. I had attributed Robert's blunder to the less educated outlook of a bygone age. But I recently found the same blunder in a popular undergraduate text, one of whose authors is reputed an authority in psychological statistics (*Introductory statistics for the behavioral sciences*, 5th ed., Welkowitz, Ewen, & Cohen, 2000, p. 260, problem 1).

22. What factors will change β without changing α?

22ANS. The direct way to answer the question, what factors affect β but not α, is to recognize that you fix α; nothing can change it (barring failure of the assumptions). Hence the three main statistical determinants of β, namely, size of real effect, sample size, and error variance are all answers to the question.

Three further comments may be of interest. (a) Increasing the real effect can't affect α because α refers only to the hypothetical case of zero effect. (b) Changing sample size actually changes the distribution of F, given H_0 true. This is why F^* depends on the df. But this F^* is selected for given df precisely to fix α at the value you specify. (c) Changing error variance has no effect on the sampling distribution of F, given H_0 true. This may seem surprising because the sample error variance is the denominator of F. But given H_0 true, the numerator of F is an (independent) estimate of the same error variance (see expected mean squares in Chapter 3). This striking feature of Fisher's F and Student's t makes them applicable without requiring any empirical information about the particular empirical situation (beyond sample size and statistical assumptions).

23. Q had a keen hypothesis that pain could be reduced by distracting attention. She measured the time subjects would keep their hand immersed in ice water until it became "too painful." A voice-actuated computer game, rock music, and a no-distraction control were tested. Her experimental procedure was impeccable, but the results were not statsig ($\alpha = .05$).

Nothing daunted, Q tried again, thinking she just had not used the right distraction conditions. She did more pilot work, talked with the subjects to understand their phenomenology of pain, and worked up a new set of distraction conditions that seemed more effective. Still not statsig.

"Third time's the charm," said persevering Q. Sure enough, her third experiment got a statsig result. "I knew my hypothesis was true," cried Q, "I just had to find the right distraction conditions!"

   a. Why do the first two failures taint the success of the third experiment?

   b. Statistically, why is Q's effective α at least .1426?

   c. What should Q do now?

   d. What is one moral of this exercise for journal policy?

23ANS. a. The first two failures taint the validity of the statsig result in the third experiment because they gave two previous opportunities for a statsig result; see also (b).

b. If H_0 is true, there is a .05 chance of statsig result in the first experiment. Under the .95 chance the first experiment is not statsig, the experiment is done again, so there is a .95 × .05 chance that the second experiment will be performed and be statsig. Similarly, there is a .95 × .95 × .05 chance that the third experiment will be performed and be statsig. Summing these three chances yields .1426. (Q's effective α level would be greater than .1426 if she had any likelihood of persevering beyond the third experiment.)

c. Q has a simple resolution to her dilemma: *Replicate*. If Q has indeed found the right distraction conditions, replication (based on a power calculation) should be successful.

d. One moral for journal policy is to emphasize replication—the universal solvent.

24. Referring to α escalation, one prominent text states "When the number of [null] hypothesis tests increases, so does the probability of a Type I error [=false alarm]" (pp. 177, 169). Explain to these authors their conceptual error.

24ANS. False alarms can occur only for a true null hypothesis. Merely increasing the number of null hypotheses that are tested, as the statement considers, does not necessarily increase the number that are true; and hence does not necessarily increase false alarms.

25. Verify the text statement that the means 2, 4, 1, 3 show zero linear trend.

25ANS. Use the given linear trend coefficients to get $-3 \times 2 - 1 \times 4 + 1 \times 1 + 3 \times 3 = 0$.

26. In the next to last paragraph on page 100, justify the assertion that "For this purpose, a familywise α seems essential."

27. Some proportion of random samples from a normal distribution will yield confidence intervals that do not contain the population mean. Can this proportion be decreased by

   a. Increasing n?   b. Decreasing MS_{error}?
   c. Increasing real effect?

27ANS. a,b,c. No. Confidence *equals* 1 – the proportion that do not contain the population mean. Of course, all three actions yield tighter, more meaningful confidence intervals.

28. Yes or no: The miss rate β should not be made as small as possible.

28ANS. No. Granted a real effect, you can make β arbitrarily small by making N arbitrarily large, but the added cost soon outweighs the added benefit.

29. Why is the 95%/99% preferable to the 99%/95% confidence interval? Why cannot this dilemma be resolved by choosing a compromise confidence level of, say, 97.5%?

30. Verify the "visual rule of thumb" in Section 4.1.2.

31. You teach the undergraduate honors course in research method, in which you explain that a statsig result is unlikely if there is no real effect. You notice, however, that many of your students mistake this to mean that a statsig result implies a real effect is likely. How do you handle this?

31ANS. That a statsig result is unlikely by chance does not warrant any *logical* inference about likelihood of a real result. This point was illustrated with the example of effect of prayer on plant growth (pp. 49f).

Empirically, however, something else is also given, namely, the belief in a real result that led the investigator to do the experiment (see *Why the Significance Test Works* in Section 19.2.2). Empirically, therefore, the inference is not illogical. Accordingly, I doubt the advisability (not to say the effectiveness) of trying to teach logical correctness. Instead, I suggest conditioning the caveat: "This too may be a false alarm."

ANSWERS FOR CHAPTER 5

1. A two-way factorial graph can be plotted in two complementary ways: With levels of the column stimuli on the horizontal or with levels of the row stimuli on the horizontal. Although both ways are equivalent, one may be more intuitive than the other.

    a. Plot the complementary form of the factorial graph of Figure 5.2. Which do you consider more useful?

    b. Plot the complementary form of the factorial graph in the left panel of Figure 5.4. Which way do you consider more useful?

    c. Do the same as in (b) for the center panel of Figure 5.4.

    d. Compare both forms of the two-way, AB data table of Figure 5.5.

1ANS. a. The complementary factorial graph for Figure 5.2 shows four data curves, one for each home cage environment. Perhaps this is preferable since each curve shows directly the mother effect, which presumably has the main interest.

b. In the complementary format for the left panel of Figure 5.4, the curves for B_1 and B_2 go in opposite directions, which may be somewhat more meaningful than the crossover.

c. The format in the center panel of Figure 5.4 seems more meaningful because the slopes of the two curves have a ratio of 9 to 3, larger than the 12 to 6 in the complementary format.

2. Piaget claimed that young children are unable to integrate two informers. Instead, they *center*, that is, they judge using only one of two given informers. Piaget considered centration to be a pervasive characteristic of cognition up to age 6 or even older. For the Intent × Damage design described in relation to Figure 1.3, draw factorial graphs, labeled appropriately, that show the following patterns of behavior:

    a. Centration on Damage.

    b. Centration on Intent.

    c. Blame = Intent + Damage.

    d. Blame = Intent × Damage.

    e. Accident-configural integration: Blame independent of amount of damage if damage is accidental; otherwise, Blame = Intent + Damage.

    f.* Centration on the larger of Intent and Damage.

2ANS. Labeling should specify one variable on the horizontal axis, the other variable as curve parameter, with the response labeled on the vertical axis.

a. To show centration on Damage, plot Damage on the horizontal. Then the curves for each level of Intent will be the same, thereby showing that Intent has no effect. The curves should not be horizontal, which would show no effect of Damage either.

    Alternatively, plot Intent on the horizontal. Then centration on Damage will appear as horizontal curves, separated vertically.

b. To show centration on Intent, interchange Intent and Damage in either graph of (a).

c. Blame = Intent + Damage. Plot either Intent or Damage on the horizontal. The curves should then be parallel. Of course,

they should be nonhorizontal and separated to show main effects of both variables. This parallelism test of the additive model is usually weak unless both variables have substantial main effects.

d. This graph should exhibit a linear fan. This is most easily shown by using numerical values for each level of each variable. The variable used on the horizontal should be plotted at its numerical values, not at equal spacing.

e. Accident-configural rule. This accident-configural rule appeared in Figure 1.3, page 27.

f*. Centration on larger of Intent and Damage. If all levels of Intent were larger than all levels of Damage, this rule would be indistinguishable from centration on Intent. And analogously for Damage.

    Useful test for this rule thus requires interlacing of the effective values of Intent and Damage. The graph will then appear as a triangularized version of the centration rules. A rule of this type has been found in taste psychophysics (McBride & Anderson, 1991).

3. Plot the two-way factorial graph for the following means.

2	11	6	7
7	17	10	12

    a. Relying on visual inspection, guess which sources in the Anova will be substantial, which will not.

    b. What features of your graph are clues to the variability of the data?

    c. What can Anova add to this visual inspection?

3ANS. a. Visual inspection indicates substantial effects of both row and column variables. Since the difference in each column is about 5, the residual will be small.

b. An (uncertain) clue to low veriability is the near-constancy of differences across columns.

c. Anova can add a proper measure of variability, based on variability *within cells*, which can give proper confidence intervals and significance tests. This table is small to rely heavily on visual inspection.

4. In the left panel of Figure 5.4, the two levels of A have quite different effects. Yet the text says ''The null hypothesis for A is true in this case.'' What's going on?

4ANS. This exercise reemphasizes that the main effect of one factor refers to averages over the other factors. What's going on is that the real effects of A average out over B in the main effect.

5. For the data table of Figure 5.2:

    a. Use the hand formulas of the Appendix to this chapter to show that $SS_{rows} = 336$ and $SS_{columns} = 228$.

    b. By visual inspection, say what $SS_{residual}$ will be.

5ANS. b. Visual inspection shows perfect parallelism so $SS_{residual}$ must be zero.

6. Do Anova for this 2×3 data table ($n = 2$) and interpret the results.

1, 2	3, 4	5, 6
1, 2	1, 2	1, 2

If you had a reasonably firm expectation that the observed data would turn out as shown in this table, what alternative analyses would you plan?

6ANS. If I had a reasonably firm expectation that the data would turn out with the given pattern, I would not do the overall Anova (except to get an overall error term). It is weakened by the expected equality of the means in the first column.

Instead, I would do a trend test on the first row and a subdesign Anova on the second row. Given that a null effect is expected, it would be important to include a confidence interval for the means of row 2 (see "*Accepting* H_0" in Section 4.1.3).

To show a main effect of the row variable, if that is necessary, I would consider a subdesign Anova on the last column because the largest A effect is expected here. Alternatively, include a subdesign Anova on the last two columns. If power is marginal, then it would be worthwhile considering which alternative is better. If power is not marginal, simplest is generally best. With $n = 2$, of course, these analyses have little power but the same principle holds for larger n.

7. For the numerical example in the Appendix to Chapter 5:

a. Verify the calculation using a computer.

b. Using Expression 11, show that the 95% confidence interval for the difference between the two row means is 3 $\pm$ 2.00. Does this agree with the F test? Which, if either, is preferable?

c. Why does the confidence interval for column means listed in the Appendix differ in length from that for row means?

d. Show that the error bar for a cell mean is ± 1.00.

e. How does this error bar relate to the standard deviation of a cell mean?

f. How does this error bar relate to the standard deviation of the population of scores?

7ANS. b. For the numerical example presented in the Appendix to Chapter 5, $MS_{error} = s^2 = 2.00$; $t^* = 2.447$ on 6 df. Accordingly, the confidence interval for the difference between the two row means is

$$5.667 - 2.667 \pm \sqrt{2} \times 2.447 \times \sqrt{2/2 \times 3} = 3.00 \pm 2.00.$$

c. Column means have a smaller "n" and hence a wider confidence interval.

d. The error bar is obtained by setting $t^* = 1$ in Expression 9 for the confidence interval. The "n" for a cell mean is 2. Hence the error bar for a cell mean is $\pm \sqrt{2/2} = \pm 1.00$.

e. The error half-bar for a cell mean equals the standard deviation for that mean.

f. The relation between the standard deviation of the population of scores and the standard deviation of a cell mean follows the law of sample size: The standard deviation of a cell mean is $\sqrt{n}$ smaller than the standard deviation of the population. The latter is estimated by $\sqrt{MS_{error}}$.

8. You need to follow up a published study with a 3×2 design that reports A means of 1, 2, 3, B means of 1.35 and 2.65, $F_A = 3.33$, $MS_{error} = 3.60$, $n = 6$.

a. First, use the hand formulas of the Appendix to check whether $F_A = 3.33$ is consistent with the other cited figures. As a first planning step, you calculate power for main effect of A in an exact replication, taking the given data at face value.

b. Show similarly that power for B is about .50.

c. What do you do about your follow-up?

8ANS. To calculate power, we need to get MS_{error} by calculating MS_A and dividing by the given F of 3.33.

a. $SS_A = 12[1^2 + 2^2 + 3^2] - 36[2^2] = 24$; $MS_A = 12$. Hence $MS_{error} = 3.60$ is consistent with $F = 3.33$.

$\sigma_A^2 = [(1 - 2)^2 + (2 - 2)^2 + (3 - 2)^2]/3 = 2/3$.

$f = \sigma_A / \sigma_e = \sqrt{2/3} / \sqrt{3.60} = .430$.

$\phi = .430 \times \sqrt{12} = 1.49$. Power for A is about .58.

b. $\sigma_B^2 = [(1.35 - 2)^2 + (2.65 - 2)^2]/2 = .65^2$.

$f = .65 / \sqrt{3.60} = .343$. $\phi = .343 \sqrt{18} = 1.46$. Power for B is about .52.

c. These power estimates are too low. Once again statsig results are not too reliable. Although larger n would yield more power, it would seem advisable to look instead for some way to reduce error variance, either through good procedure, or through design, such as repeated measures.

9. In the next to last paragraph under *Advantages of Factorial Design*, evaluate Q's suggestion that P replicate her 2×2 design twice is "better in every way"? List Sources and df for this Anova.

9ANS. The main advantage of Q's suggestion is that it includes P's two replications as a factor in the design. It is thus possible to assess whether similar results are obtained in both—with no more subjects than P actually used. Also, of course, all confidence intervals would be shorter because the error term would be based on more df from the replicated experiment. With this form of replication, P would gain the greater efficiency of Q's design. Replications is a third factor, so the design becomes a 2^3 factorial—three main effects and four residuals.

10. In a survey of his work on interpersonal attraction, Aronson (1969, p. 159) presented two hypotheses derived from a rationale of anxiety reduction:

Hypothesis 1: "If a beautiful woman were to evaluate a male subject favorably, she would be liked better than a homely woman who evaluated him favorably."

Hypothesis 2: "If a beautiful woman were to evaluate him unfavorably, she would be liked *less* than a homely woman who evaluated him unfavorably."

Aronson concluded "The results confirmed our predictions: There was a significant interaction—the beautiful-positive evaluator was liked best, but the beautiful-negative evaluator was liked least" (p. 161).

These quotes refer to an experiment by Sigall and Aronson (1969), in which male subjects were told that a first-year graduate student in clinical psychology would test them as part of a department-wide program. This person was actually a personable female undergraduate, blind to the purpose of the experiment, made up to appear attractive for a random half of the male subjects, unattractive for the other half. She gave each subject the California Psychological Inventory, spent seven minutes in an alleged "examination" of the completed form, and gave the subject a favorable or unfavorable evaluation of his personality according to a predetermined schedule, regardless of his answers. She then left and the experimenter obtained the subject's responses to some questions on a −5 to +5 scale that were emphasized to be totally anonymous.

Mean ratings of liking for the confederate were ($n = 12$):

Female	Attractive	Unattractive
Favorable	3.67	1.42
Unfavorable	1.08	1.17

For the interaction, $F(1, 44) = 7.87$.

    a. Are the two quoted hypotheses of equal interest?

    b. Does visual inspection of the data table confirm both quoted hypotheses?

    c. Does the significant interaction confirm the quoted hypotheses?

    d. How did the authors err in doing the factorial Anova?

    e. Should this experiment have been published?

10ANS. a. No. Hypothesis 1 is obvious. Hypothesis 2 is interesting.

b. Visual inspection supports Hypothesis 1 (since 3.67 looks considerably greater than 1.42) but not Hypothesis 2 (since 1.08 is surely not statsig less than 1.17).

c. No. The statsig interaction did not confirm their main prediction, Hypothesis 2. In particular, statsig interaction can be obtained if the two means in the second row are equal—as in these data.

d. The authors stated two specific hypotheses, as quoted. Each is a two-mean comparison (Chapter 4) of the two means of one row of the data table. The only appropriate analysis is to test both two-mean comparisons. That the conditions formed a factorial design is entirely irrelevant. The analysis should address the theoretical hypotheses.

e. Publication does not seem warranted. The one result is trivial. Replication would seem mandatory before presenting this outcome as bona fide.

    Indeed, the contrary result was obtained with a second response measure. This second question asked how many additional sessions the subject would be willing to work with the confederate in future studies of this testing program. Mean number of additional sessions was 2.83 for the attractive-unfavorable condition, 1.08 for the unattractive-unfavorable condition, which would evidently have been comfortably statsig.

    This contradiction was hand-waved away by saying that this response tapped "different needs."

    This contradiction between two response measures also points up the danger of depending on only one, as many experimenters do.

10a. In the Sigall–Aronson data of the preceding exercise, show that the 95% confidence interval for Hypothesis 2 is .09 ± 1.19. What does this mean?

11. Using the three rules for degrees of freedom, show that the df for all systematic sources, including mean, add up to the total number of treatment means:

    a. for a two-way design;   b. for a three-way design.

12. a. Plot a two-way factorial graph for the following means. Relying only on visual inspection, give your judgment about which sources are substantial, which are not.

    b. What features of the figure are clues to the variability of the data?

    c. What can Anova add to this visual inspection?

7	21	36
14	23	30
19	24	26

12a. ANS. All three row curves show a steady upward slope. Although these slopes differ, each curve exhibits a roughly linear shape with equal spacing of the column variable. This replicated pattern suggests a substantial real effect of the column stimuli.

    The large difference in slopes of these curves indicates that the effect of columns depends strongly on the row level. Hence the residual $MS_{row \times column}$ should be substantial.

    Despite this slope difference, however, the means of the three row curves, which are roughly equal to their middle points, look nearly equal. The main effect of rows thus appears small.

    b. The main clue to reliability in these means comes from the overall left-to-right increase of the column effect and its steady decrease down successive rows, accentuated by the relatively small pointwise departures from this pattern. Note that three row curves are considerably stronger evidence that two would have been.

    If the row and/or column levels were empirically known to be ordered in some way that would imply such regular change in response, this extrastatistical knowledge would reinforce the foregoing face interpretation of the data.

    c. Anova can add a measure of chance error based on *within* cell variability. This error measure is more general and more valid than can be obtained from the graph of means. This error term can provide confidence intervals and significance tests that are free of the observer's biases and chance patterns that catch the observer's eye. This is valuable in situations in which the factorial pattern is less definite than in this artificial example and especially with 2×2 and other small designs.

13. You have a 2×3, $A \times B$ design with entries of 12, 18, 25 in the first row. Construct entries for the second row such that the only nonzero sources are:

    a. B.   b. AB residual.   c. B and AB residual.

    d. Give a verbal argument to show that it is not possible to find a second row such that only A is nonzero.

    e. Prove (d) algebraically.

13ANS. d. Since a column effect is already present in the first row, the only way to get zero B effect, that is, equal column sums, is with a second row opposite in trend to the first row. But this implies nonparallel row curves, hence nonzero AB residual.

e. To have AB zero requires the two rows to be parallel. The second row must thus have the form $11 + c$, $18 + c$, $25 + c$; and c must be nonzero to get A nonzero. This does yield zero AB. But then the column sums cannot be equal so B is not zero.

14. Above Equation 3a, the text asserts each SS is a sum of squared deviations.

   a. Explain what these deviations are in Equation 3a.

   b. What is the substantive significance of these deviations?

   c. Do the same for Equation 3e.

   d.* If the factorial plot is parallel, show that all deviations in Equation 3ab are zero.

14ANS. a. The deviations in Equation 3a are the differences between the row means and the overall mean.

b. Each given deviation is the *relative* mean effect of a row treatment, that is, $\hat{\alpha}_j$. The size of these deviations is our evidence for real effects.

c. In Equation 3e, each score in parentheses is the difference between the response of some individual subject and the mean score for all the subjects in the same experimental condition. These differences thus reflect individual differences between subjects treated alike. This is just what we need for our yardstick of error variability.

15. In the last paragraph of Section 5.2.1, justify and/or fault the statement that visual inspection of the four means in a 2×2 factorial graph "can hardly assess reliability of main effects."

15ANS. A natural reservation with this statement is that the top curve might seem visually "far away" from the bottom curve in the 2×2 graph. But without any knowledge of error variability, statistical or extrastatistical, "far away" has no meaning.

   If you have some kind of background information about the error variability, of course, or about the expected size of the effects, then you have some ground for judging "far away." The statement may thus be faultable in that visual inspection will—and should—incorporate background knowledge.

16. In a 4×3 design, suppose the A values are 1, 2, 4, and 8, and the B values are 0, 1, and 3. Assume an additive model so that each cell entry equals the sum of the corresponding values of A and B.

   a. Plot the factorial graph. Prove that the curves are parallel.

   b. Do (a) with A values of α_1, α_2, α_3, α_4, and B values of β_1, β_2, β_3.

16ANS. By the additive model, the mean for cells jk and $j'k$ are $(\alpha_j + \beta_k)$ and $(\alpha_{j'} + \beta_k)$. The difference is $\alpha_j - \alpha_{j'}$, which is constant across columns. QED.

17. You are TA in the undergraduate honors class in animal behavior. In the term paper experiment, one of your students writes: "My obtained F was greater than the criterial F^* for $\alpha = .05$. Therefore, I may reject the null hypothesis and conclude that the population means are significantly different. I realize,

of course, that this conclusion has a 5% chance of being wrong." Grade this answer constructively.

17ANS. Congratulations on getting reliable results in your interesting project, which is often difficult in the short time available in this class. Two somewhat subtle statistical points deserve your attention. First, "significantly different" can only apply to the sample means; you refer to the population means, which are either equal or unequal. The sample means are clearly unequal; the question is whether their difference is large enough to give reasonable confidence that the population means are unequal.

   Second, the significance level does not quantify the probability of being wrong. The 5% chance is *conditional* on H_0 being true; but in that case the conclusion has a 100% chance of being wrong. If H_0 is false, on the other hand, the conclusion has 0% chance of being wrong. These statisticalities aside, you have a fine project—an auspicious sign for your life path.

18. Sometimes you need the error term from published data to get a confidence interval not presented in the article, for example, or to make a power calculation. In the numerical example in the Appendix, suppose you were given the row means, 2.667 and 5.667, each on 6 scores, together with their F of 13.50.

   a. Get MS_{error} from this given information.

   b. Estimate power for the row effect in an exact replication.

18ANS. a.
$$SS_{rows} = 6[2.667^2 + 5.667^2] - 12 \times 4.167^2 = 27 = MS_{rows}.$$
$$MS_{error} = MS_{rows}/F = 27/13.50 = 2.00.$$

b. For the power calculation for row effect, we have $\sigma_\varepsilon = \sqrt{2}$. With $\bar{\mu} = 4.167$, $\alpha_1 = 2.667 - 4.167 = -1.500$, and $\alpha_2 = +1.500$.
Hence $\sigma_A = \sqrt{(1.5^2 + 1.5^2)/2}$, visual inspection of which shows that $\sigma_A = 1.5$.
The standardized power effect size is thus
$f = \sigma_A/\sigma_\varepsilon = 1.5/\sqrt{2} = 1.06.$
Thus, $\phi = 1.06 \sqrt{n}$. With $n = bn = 6$, $\phi = 2.60$. From the power graph for 1/6 df, power is about .86.

19. Construct data for a $2 \times 3 \times 2$ design with all effects zero except:

   a. A and B.   b. BC.   c. A and AB.

(It may help to sketch a factorial graph, then translate it into numbers.)

19ANS. b. A simple solution is to use -1 and $+1$ in the first row of the BC table, 0 and 0 in the second row, and $+1$ and -1 in the third row. Then BC is nonzero, but the pure crossover makes both marginal means zero. Duplicate this for both levels of A.

c. Rotate the BC table of (b) into a 2×3 design and relabel the variables A and B. Add a constant to the first row, making A nonzero. Use for both levels of C.

20. With $n = 2$, Figure 5.5 is based on 48 scores. Each score is one unit of information, so the total df is 48. Where do these 48 df appear in the Anova?

20ANS. Each cell in Table 5.5 has 2 scores, so each cell provides 1 df for error, for a total of 24, and 1 df for systematic sources, also for a total of 24. The systematic sources are Mean with 1 df, A with 2 df, B with 3 df, C with 1 df, AB with 6 df, AC with 2 df, BC with 3 df, and ABC with 6 df, for a total of 24.

21. a. Make up a realistic set of data for the 2×2, praise–blame experiment of Note 5.2.2a that satisfies the indicated pattern.

b. Introvert/extrovert was not an experimentally controlled variable in this experiment. Instead, it was a score on a personality test. Perhaps the effect is due not to personality, but to some confounding characteristic of the subjects that is correlated with the personality test. Discuss whether confounding with motor skill seems a likely explanation of these results. What data did the investigators measure, not included in Note 5.2.2a, that would bear on this issue of alternative explanations?

21ANS. a. A realistic set of data from the praise–blame experiment should show a large crossover with small main effects. A cancellation rate of perhaps ½ to 1 digit per second might be expected for fifth graders. The actual data for the later trials were:

Extroverts praised	25	Extroverts blamed	31
Introverts praised	30	Introverts blamed	24

b. If subject differences in motor skill were causal in this situation, then the extroverts should be better or poorer under both reinforcement conditions, contrary to the crossover.

Stronger evidence comes from the first trial, before any reinforcement had been given. On this trial, the original report showed that all groups were nearly equal, which argues against the possibility that the two personality groups differed in motor skills. (Further support comes from the near-equality of the four subgroups on the second trial.)

It is instructive to add that performance on the first trial was less than 8 digits, but nearly 26 on the second trial. Presumably the children were confused about the task on the first trial. This result indicates the importance of initial practice.

22. In your experiment, based on a 3×4 design, with $n = 3$, you have an a priori theoretical hypothesis that all cells with A fixed at A_1 have equal true mean. How do you use Anova to test this?

22ANS. Your theoretical hypothesis refers to the four true means for the first row of the design. Do one-way Anova on this subdesign, but use overall MS_{error} on 24 df.

23. In the previous exercise, you have an a priori hypothesis that all cell entries with B fixed at B_1 are zero. How do you test this?

23ANS. Do Anova on the subdesign of the three cells in the first column of the design. This provides two tests, both necessary. First, F_{mean} should be nonstatsig. Second, F_A should also be nonstatsig. Treat MS_{error} as above.

24. For power calculations for three-factor design (Section 5.2.5) with n subjects in each cell, what are the values of "n" for the three main effects? For the AB residual?

24ANS. The values of "n" for A, B, C, and AB are bcn, acn, abn, and cn.

25. People construct situation models that integrate specific information from the situation at hand with nonspecific world knowledge to use for goal-directed action in the situation. How to analyze the structure of such situation models is a central theoretical problem. With spatial situations, such mental models involve priming gradients whose structure can be studied using response times.

In the following experiment, subjects thoroughly memorized a diagram of a rectangular laboratory building with 10 rooms, each containing two to four named objects. Next, they read a narrative of some 20 sentences that served to motivate the situation confronting a protagonist in the laboratory; this was followed by a motion sentence in which the protagonist moved from a *start room* to a *goal room* (e.g., "Then he walked from the storage room into the lounge"). These motion sentences required the protagonist to move through one or two unmentioned *path rooms* between start and goal rooms.

Finally, subjects read a probe sentence that referred concretely to a memorized object in one of the one or two unmentioned path rooms through which the protagonist would have had to move to get to the goal room from the start room.

The response measure was the reading time for this probe sentence.

Previous work had established that reading time depended on the location of the object along the protagonist's path; the motion sentence induced a priming gradient such that reading time increased steadily with distance *backward* from the goal room in which the protagonist had finished his motion.

Two hypotheses about the distance gradient were tested. Under the standard *Euclidean distance hypothesis*, the distance was the actual physical distance. Under the *categorical distance hypothesis*, the distance would depend on the number of rooms, but not on the physical distance, contrary to the Euclidean hypothesis. Previous studies had confounded these two variables. The present study deconfounded them.

Three variables were manipulated in a 2^3 design relating to the room in which the object in the probe sentence was located. *Room size* (short, long) was expected to have an effect under the Euclidean distance hypothesis, no effect under the categorical distance hypothesis. *Room division* (divided, undivided) was expected to have an effect under the categorical distance hypothesis, no effect under the Euclidean distance hypothesis. *Near-far* referred to the Euclidean distance between the goal room and the object, which was expected to have no effect under the categorical distance hypothesis.

a. Evaluate the two hypotheses by visual inspection of Figure 5.7.

b. What can Anova add to this visual inspection?

25ANS. a. The graph shows negligible difference between short and long room sizes. This strongly disagrees with the Euclidean distance hypothesis and mildly supports the categorical distance hypothesis.

Note especially the near identity of the four pairs of data points for this room size variable. This gives a visual index of error variability that underlies the visual inspection.

With this visual index of error, divided rooms clearly took longer time, in line with the hypothesis of categorical distance, contrary to the hypothesis of Euclidean distance.

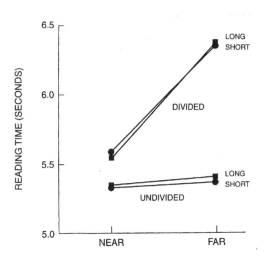

Figure 5.7. Reading times in experiment on organization of spatial mental models. (After Rinck, Hähnel, Bower, & Glowalla, 1997.)

Of special interest are the bottom two lines of Figure 5.7, which show explicitly that Euclidean distance had negligible effect when the room was undivided. When the room was divided, then the path distance greatly amplified the room division effect, as shown by the large upward slope of the top two lines.

b. Anova adds little to visual inspection. The reason is that a visual index of error variability is given by the negligible effect of room size; all four pairs of data points are close together. Beyond indicating that room size has negligible effect, the differences within pairs is a measure of error.

However, the small divided-undivided difference for the near path seems to have theoretical significance, for which a two-mean comparison would be needed. For this reason, among others, an appropriate within-condition measure of error variability should be included, as by including a confidence interval in the graph (see Exercise 6.9).

The importance of the room size variable for this visual inspection deserves emphasis. Without this variable to give a visual index of error, the 2 × 2 design is too small to be amenable to visual inspection. As a methodological tip, inclusion of a minor variable for balance, such as two or more experimenters, could be used in the same way, and may usefully be included in a figure as was done by Rinck, et al.

26. You have a $3 \times 4 \times 2$ design, with $n = 5$. a. Write out Source and df for the Anova table. b. You have a theoretical hypothesis that all cells with A fixed at A_2 and with C fixed at C_1 have equal true means. How do you test this?

26ANS. a. The sources with their df are:
Mean, 1; A, 2; B, 3; C, 1; AB, 6; AC, 2; BC, 3; ABC, 6; error, 96.

This exercise aims to help students internalize the pattern for calculating df. This pattern is useful in checking computer output, even more in checking reports in the literature.

b. This refers to a one-way subdesign with only B varied. Test as in Exercise 22.

27. This exercise has two goals. One is practice in estimating the main effects, α_j, β_k, and the residuals, $(\alpha\beta)_{jk}$, from factorial data. More important is to illustrate that the residuals may lack psychological meaning (see further Exercise 7.5).

a. For the following 3×4 table of cell means, use visual inspection to localize the interaction residual. Assume error-free data.

1	2	3	4
3	4	5	6
6	7	8	3

b. Show that $\alpha_j = -1.83$, .17, and 1.67, and that $\beta_k = -1.00$, .00, 1.00, and .00, using Equation 1b.

c. Use the expression for $(\alpha\beta)_{jk}$ in Equation 1b to show that all 12 Anova residuals are nonzero. What is the moral of this demonstration?

ANSWERS FOR CHAPTER 6

1. You are TA in the undergraduate honors class on research method. The instructor asks you for a two-paragraph writeup to be given to the class to show advantages and disadvantages of repeated measures design. What do you write?

1ANS. Two advantages of repeated measures design noted in the introduction to this chapter are deconfounding of treatment conditions from the main component of individual differences and lower error variability. A disadvantage is possible confounding from order effects.

This exercise also bears on the two distinct concepts of between subjects variability and within subjects variability, the relation between them, and how the repeated measures design provides its advantages. That was one purpose of Table 6.1 and Figure 6.1.

2. In your undergraduate class on research method, one pair of students worked together to develop person descriptions consisting of a photograph of a male college student (the row factor) and a pair of his personality traits (the column factor) for a study on person cognition. Each factor is varied across three levels: low, medium, and high attractiveness.

Female college students judge how much they would be interested in a date with each of these males. Both students perform the same experiment using the same person descriptions, but each runs a separate group of 16 subjects from the standard pool. Each subject judges all nine person descriptions twice in randomized order. Their mean responses are as follows.

Stud.-1	11	25	45
	17	34	51
	24	38	55
Stud.-2	7	21	53
	25	30	47
	20	46	51

a. Plot the factorial graph for each student. In what way do the two sets of data agree? In what way do they disagree?

b. The two graphs presumably should show the same pattern. Hence the cited disagreement raises suspicion about one or the other set of data.

   (i) What is suspicious about the data of Student 2, and why?
   What feature of the Anova would shed light on this question?

   (ii) Argue instead that Student 2 shows the true picture.
   What feature of the Anova would shed light on this question?

2ANS. a. The two graphs agree quite well on the main effects. However, the factorial pattern seems a lot more regular for Student 1.

b(i). The three crossovers in the data of Student 2 seem suspicious. They might stem from careless procedure or casual treatment of the subjects, as seems not unlikely with beginning students in an undergraduate class.

If this suggestion is correct, the data of Student 2 should show larger error variability than for Student 1. To check this, compare MS_{error} for the two students.

Within subject variability can also be compared between the two students because each subject judged two replications.

b(ii). With person descriptions, there might well be interactions between the photograph and the trait adjectives. The data of Student 1, being near-parallel, say otherwise. But perhaps Student 2 made the task more real to the subjects and this greater reality produced the photo-trait interaction.

If this is true, then Anova of Student 2's data should show a photo-trait interaction residual.

3. a. Show that $F_A = 7.50$ for the data of Table 6.1 by hand calculation.

   b. What does this tell you?

3ANS. b. The F of 7.50 is greater than the criterial $F^*(3, 6) = 4.76$. This is reasonable evidence that the true means of the four treatments are not all equal. Regrettably, this tells you little to help localize these differences.

4. Assume sphericity for the data of Table 6.1. Using your analysis from the previous exercise, get a confidence interval for difference between A_1 and A_2.

4ANS. On the sphericity assumption, we may use $MS_{SA} = 4/6$ from the overall Anova as the error term on 6 df for confidence intervals. Hence we get $3 \pm \sqrt{2}\, t^*(6) \sqrt{MS_{SA}/3}$. With $MS_{SA} = .667$, the confidence interval is 3 ± 1.63.

5. Intuitive physics of 5-year-olds is shown in the following table for two tasks.

Right panel shows judgments of amount of liquid in glasses as a function of the height of liquid and diameter of glass (cm).

Left panel shows judgments by same 5-year-olds of area of rectangles as a function of height and width (cm). Equivalent response scale in both tasks.

   a. Graph these data and get the main implications by visual inspection. See Exercise 2 of Chapter 5.

   b. Suppose a $2 \times 3 \times 3$, Task $\times$ Height $\times$ Width Anova is run on these data (treating diameter of glass as width). By visual inspection, aided as needed by pencil and paper, say whether each Anova source would have an F ratio that is large, medium, or small (near 1).

   c. Parts a–d of Exercise 5.2 suggest four possible hypotheses for this task of intuitive physics. How can Anova test them?

Rectangle height

Width	07	09	11
11	11.4	14.7	17.7
9	7.4	11.7	13.9
7	4.2	7.2	10.4

Glass height

Diam.	2.5	5.0	7.5
8.5	3.1	10.2	16.1
7.5	4.2	11.8	17.3
6.5	4.2	12.4	16.7

(Data after Anderson & Cuneo, 1978, Figure 1.)

5ANS. a. Visual inspection shows a clear pattern of judgment on height alone for the glass task. The near-parallelism of the rectangle judgments implicates a Height + Width pattern for judgments of area. The physically correct integration rule is Height × Width, so the area data disagree sharply with expectation and hence come under question.

b. The strong upward slope of the curves in both panels implies that height will have a large F. (The main effect of height may be graphed as a curve with three points, one for each height, each the mean of the corresponding six data points in the figure.)

Since the slope of the curves is somewhat greater for liquid than for area, the task-height interaction residual will have a medium F. (The task-height interaction may be graphed as two curves, each the average of the three curves for that task in the above table. The interaction is shown by the amount of non-parallelism.)

The effect of width is substantial with area, but nil with liquid. The main effect of width in the Anova is an average of these two, so it will have a medium F.

The task-width interaction residual assesses whether the width factor has different effects in the two tasks. Clearly it does, so this source will have a medium F.

The main effect of the task variable assesses the difference between the mean of all the data points in the left panel and the corresponding mean in the right panel. This looks a bit higher for liquids, but should have a small F.

The height-width interaction residual asks about parallelism in the data averaged over the two tasks. Since each panel shows parallel data, so will their average. Hence this source will have a small F.

The three-way interaction residual asks whether the amount of nonparallelism differs for the two tasks. Since the data are near-parallel for each task, the answer is evidently no. Hence this source will have a small F.

c. Piaget's centration rule predicts that only one factor will have a main effect. The Height × Width rule predicts two main effects and an interaction. The Height + Width rule predicts two main effects, but no interaction. Anova provides straightforward tests of all these predictions.

Since the Height + Width rule was unexpected, these data only allow a post hoc interpretation, one that would need to be verified with replication. More serious than this reliability problem, of course, is the validity problem of possible shortcomings

in task–procedure. Assessing validity took seven replication experiments, in the course of which the additive rule was amply confirmed. This additive rule is thought to be one manifestation of a general purpose rule that appears also in other physical situations that obey a physical multiplication rule (e.g., Figure 20.1).

6. What specific experimental procedures would you use to reduce each of the two components of the error term, MS_{SA}, in Equation 3d in a repeated measures study of (a) rats learning to run a straight runway for food reward and (b) 7-year-old children in the blame study of Figure 1.3 of Chapter 1?

6ANS. One component of error is σ_ϵ^2, that is, variability within subjects; the other is σ_{SA}^2, which is part of the variability between subjects. Adapting subjects to the experimental procedure seems likely to reduce both components. With rats, a preliminary period of gentling to adjust them to being handled is often appropriate. With children, ask a few questions about what they like and dislike to help them feel at ease.

7. Some people live dismal half-lives because they suffer never-ending, chronic pain that little can be done about. You hope to show that your combined use of hypnosis plus drug is superior to either hypnosis or drug alone. You use a repeated measures design, with volunteers from a pain clinic whom you have screened to be reasonably hypnotizable. You are successful: The combined treatment is superior to either one alone, with a statsig F and a respectable effect size.

   a. What may you conclude about the effect on individual subjects?

   b. How can you amplify your design to get firmer evidence on individuals?

   c. Is it appropriate to screen subjects as indicated?

7ANS. a. You may conclude that hypnosis plus drug is superior for some individuals, although you may not be completely sure which ones these are. You may not conclude that hypnosis plus drug is superior for all individuals. Indeed, extrastatistical knowledge suggests your treatment probably has little value for a fair fraction of the individuals.

b. By replicating for each subject, you could get a measure of variability for each individual. Single subject analysis is then possible (Chapter 11).

c. Screening to select reasonably hypnotizable subjects seems only sensible. Chronic pain is a terrible affliction; any relief for even a minority of patients is worthwhile.

8. You have an $(S \times A) \times G$ design, with $n = 8$, $a = 4$, and $g = 5$.

   a. Write out Sources and numerical values of df. Check that the sum of your df is correct.

   b. How can you get a confidence interval for G_1 versus G_2?

   c. How can you get a confidence interval for A_1 versus A_2?

8ANS. a. The sources and df are:
Mean, 1; A, 3; G, 4; AG, 12;
S/G, 35; SA/G, 105.

Check: $\Sigma\,df = 8 \times 4 \times 5$.

S/G is the error term for Mean and G; SA/G is the error term for A and AG.

b. The Anova in the Between subtable is on a single score for each subject so the methods of Chapters 3 and 4 apply. Assuming equal variance, $MS_{S/G}$ from the Between subtable is the MS_{error} for the confidence interval between any two group means.

c. The within error, SA/G, is not generally appropriate to test between A_1 and A_2 because of likely nonsphericity. This test needs its own error term. Accordingly, give the data for just these two levels of A to the computer. This subdesign Anova yields F_A for these two levels on 1/35 df. A confidence interval may be obtained using the within error from this subdesign Anova, with t^* on 35 df.

9. In the experiment of Figure 5.7 in Exercise 25 of Chapter 5, each subject went through all 2^3 experimental conditions, a repeated measures design.

　　a. Although small, the difference between divided and undivided path rooms for the near path looks perhaps larger than chance. How would you get a confidence interval to test this difference?

　　b. Suppose you obtain a similar confidence interval for the far path. How do you think its width would compare with that of (a)?

9ANS. a. Each subject has four data points for the near path. Give these four data points to the computer as a 2×2 repeated measures Anova (with three error terms). MS_{error} for the confidence interval is the error term for divided-undivided in this Anova, to be used in conjunction with Expression 1 for confidence intervals of Chapter 4.

b. The corresponding confidence interval for the far path should be wider because individual differences will presumably be greater for a larger effect. Part of these individual differences goes into the error mean square for the confidence interval.

10. You are writing up the experiment of Table 6.2 for publication. You notice that the F of 2.00 in Table 6.2 is close to statsig at $\alpha = .05$. How do you handle this? Indicate whether follow-up tests on these data might be useful.

10ANS. One student gave this cogent answer (a little revised):

Inspect the factorial graph. An effect localized in 1 of the 6 interaction df could be sizable but diluted by the other 5 df, thereby making the overall residual nonstatsig. Such an effect could occur as a fanning pattern in the factorial graph, corresponding to the 1 df for the linear × linear component of the interaction.

I may supplement this answer by noting that the next step depends on the outcome of the visual inspection. If no likely meaningful locus appears, it would suffice to note in passing that "The AB residual was close to statsig, but inspection of the factorial graph revealed nothing further."

On the other hand, suppose a likely locus is found:

　(i) If the presumed effect looks important, do a post hoc significance test. If this is statsig, plan a replication experiment to verify it.

　(ii) If the presumed effect does not look important but is still something other investigators should keep in mind, do a post hoc test and report it as a post hoc test.

11. On page 163, the text states that "σ_A^2 will be greater than 0, [and] MS_A will tend to be larger than MS_{SA}." Why does the first clause assert "will be" while the second clause says only "will tend to be"?

11ANS. The question revolves around the difference between population and sample. The premise is that there are real effects, that some α_j are nonzero in the Anova model for the population. Hence $\sigma_A^2 = \Sigma\,\alpha_j^2/a$ is also nonzero.

In contrast, MS_A and MS_{SA} refer to the sample. Given real effects, the former is larger than the latter on average, but not in every sample.

12. You are in your second year as assistant professor, busily pursuing the program of research you began with your Ph.D. thesis. Glancing at the latest issue of the main journal in your field, you notice an article that seems to blow your research program to the moon. Your first thought is to replicate this experiment, adding what you consider an essential control. Your first step for replication is a power calculation. The report gives means of 5 and 1 for the two critical conditions, each with 32 subjects, and an F of 4.00.

　　a. From the given data, show that an exact replication would yield power about .50.

　　b. Now what do you do?

12ANS. a. Means of 5 and 1 yield $\alpha_1 = (5 - 3) = 2$, and $\alpha_2 = (1 - 3) = -2$. Hence $\sigma_A = \sqrt{(\alpha_1^2 + \alpha_2^2)/2} = 2$.

To estimate σ_ε, calculate $SS_A = 256 = MS_A$. Divide by $F = 4.00$ to get $MS_{error} = 64$; $\sigma_\varepsilon = 8$.
(This F is based on the repeated measures error.)

Then $\phi = \sigma_A/\sigma_\varepsilon \times \sqrt{n} = (2/8)\sqrt{32} = 1.4$. The power chart shows power just a bit over .50. This power is much too low to warrant an exact replication.

b. A larger n would yield more power, but n of 32 is already pretty large. One possibility is to find some confound in the article that protects your thesis project. Alternatively, you could smile through your tears for timely prevention from an unrewarding project.

13. In Section 6.4.3, why does a confidence interval for a single mean require a between subjects error whereas a confidence interval for the difference between two means uses a within subjects error?

13ANS. In taking the difference between two means, the main effect of subject is subtracted out. In contrast, the variability of a single sample mean consists mainly of between subject differences. With only a single mean, these individual differences are not subtracted out. The confidence interval must allow for the operative variability.

14. There is a numerical relation between the df in the Between subtable and the df in the Within subtable in the $(S \times A) \times G$ design of Table 6.3.

    a. Can you discover it?

    b. Can you explain it?

14ANS. a. The df for each source in the within part of the table equal $(a - 1)$ times the df at the corresponding location in the between part of the table:

    A on $(a - 1)$ df corresponds to Mean on 1 df;

    AG on $(a - 1)(g - 1)$ df corresponds to G on $(g - 1)$ df;

    and SA/G on $g(n - 1)(a - 1)$ df corresponds to S/G on $g(n - 1)$ df.

b. The reason for this numerical relation is that each of the a levels of the within subjects variable A constitutes a replication of the design in the between subjects part of the table. Of these a replications, 1 df goes to analysis of the mean of the a scores for each subject, which appear in the Between subtable, and $(a - 1)$ remain for differences between the A means, which appear in the Within subtable.

15. Yuval Wolf (2001) presented a version of the blame experiment of Figure 1.3 to 20 juvenile delinquents, half of whom had records of violence, half of whom did not. He hypothesized that the violent subgroup would place more weight on harm, less weight on intent than the nonviolent subgroup.

    a. Draw a factorial graph showing the hypothesized pattern.

    b. How should this hypothesis appear in the Anova?

15ANS. Consider the 3×3, Intent $\times$ Harm design. The three row curves for Intent should be steeper for the violent subgroup, thereby showing greater effect of Harm, and closer together, thereby showing lesser effect of Intent. This pattern should appear as a three-way interaction, Intent $\times$ Harm $\times$ Violent-nonviolent.

16. The ab conditions of an $S \times A \times B$ design are presented once to each subject in a separate randomized order for each subject, and then a second time. This replication (R) factor may be added as another factor in the design.

    a. Write down the Anova source table, indicating appropriate error terms.

    b. Suppose the two replications are entirely equivalent, with no order effects, differing only from within subject variability. What does this mean for the terms involving R and S?

16ANS. a. Replications makes a fourth factor, so there are 15 sources. The error for each of the seven sources involving A, B, R is of course its interactions with subjects.

b. Given that the two responses from each subject are the same except for individual response unreliability, σ_e^2, the expected mean square for every source involving R also equals σ_e^2. Hence these sources can be pooled by adding their SSs and df to get a pooled error on abn df, which could be used to test all sources involving S. These individual differences might be of interest. (These F ratios assume equal variability for all subjects, which is hardly likely but which should not make much difference.)

17. Q gave preliminary practice to avoid practice effects in her repeated measures design, but she still planned to avoid possible confounding by balancing treatments across serial position. Alas, her research assistant gave all the treatments in the same order.

    a. This use of only one order of presentation does not invalidate the Anova test of the observed means. Why not? What does it invalidate?

    b. What results, if any, might be salvageable?

17ANS. a. Anova tests the null hypothesis that the true means underlying the observed means are equal. These observed means represent the joint effect of treatments *and* order, which are confounded. The Anova cannot know anything about this confounding; this is an extrastatistical matter.

Of course, this confounding threatens the *interpretation* of the data.

b. Since improvement is expected, a poorer performance on a later treatment would suggest it really was poorer. Also, if prior knowledge indicates small practice effects, large treatment differences would seem meaningful. The results may thus be useful guides to further work.

18. (Transposed from Fisher, 1958, p. 126.) The world-famous San Diego Zoo seeks to create natural surroundings for animals from around the world. Breeding and rearing these animals is one of its specialities, but it faces many problems because each species generally has unique dietary needs about which little may be known. Diet experiments are difficult with some species, moreover, because only a few animals may be available to test different diets.

Consider an animal that usually has two offspring at a time. Diets D_1 and D_2 are randomly assigned to the two members of the first birth at the zoo. The weight gain of the infant receiving diet D_2 is 90 g greater. This experiment is repeated some months later when the second birth occurs. This time diet D_2 has an 80 g advantage.

    a. Is the effect statsig?

    b. Suppose the second replication had yielded a 110 g advantage for D_2, so its mean advantage is almost 20% larger. How much does Anova say this larger effect increases your confidence? Explain.

    c. What possibly important variable was omitted in this design? How would you handle it?

18ANS. a. Yes; $F_A = 289$; $F^*(1, 1) = 161$. The main interest of this exercise is that a sample of size 2 is (just) enough to estimate mean and variance—enough to represent the mean as an interval of uncertainty. These estimates are not very reliable, of course, and Anova is sensitive to violations of equinormality with very small samples. Still, such results would encourage continued trials with diet D_2.

b. Although the mean effect is nearly 20% larger, the error variance is 400% larger. This comparison emphasizes the high importance of reducing error variance with small samples.

c. Birth order is omitted. I would want to balance diet across birth order.

19. The *additive-factor method* introduced by Sternberg (1969, 1998) seeks to dissect a response into successive, independent stages. The overall reaction time (RT) will then be the sum of the separate RTs for successive stages. If successful, this method can give unique insight on serial organization of responses.

Consider a simple task thought to consist of two stages. Seek two variables, each of which you hope will influence the duration of one stage, with no influence on the duration of the other stage. Manipulate these two variables in factorial design. If all goes as hoped, the durations of the two stages will be additive. This additivity will be observable as parallelism in the factorial graph.

One such experiment (Sternberg, 1969, Exp. V) used three variables, each at two levels to obtain a $2 \times 2 \times 2$ repeated measures design with $n = 5$. The stimuli were numerals presented visually and the response was a spoken numeral. The three variables were:

Stimulus Quality: Numerals presented *intact* or *degraded* with visual *noise*.

Stimulus–Response (S–R) Mapping: Response was the numeral itself or the numeral plus 1.

Number of Alternatives: Number of stimuli was two or eight, and similarly for the number of responses.

a. On intuitive grounds, two of these variables may be expected to affect separate stages and so have additive effects on RT. Which are they?

b. One variable may be expected to affect both stages in (a). How so?

c. Each subject received extensive preliminary training and then was given seven experimental sessions. The following table gives the means for this experiment. Plot them in a graph with two levels of one variable on the horizontal, thus obtaining a graph with four curves, representing two 2×2 designs. Of the three possible choices of which variable to use on the horizontal axis, which do you think is most informative and why?

d. Use visual inspection to interpret Sternberg's mean RTs in terms of his theoretical formulation.

e. For any 2×2 design, the deviation from parallelism may be expressed as a single number, namely, the double difference (interaction residual)

$$(\bar{Y}_{11} - \bar{Y}_{21}) - (\bar{Y}_{12} - \bar{Y}_{22}).$$

Show that this double difference is 2.2 ms for the Stimulus Quality × S–R Mapping design, averaged over Number of Alternatives.

f. The standard deviation (error half-bar) for the deviation from parallelism of (e) was 3.5 ms. Show that the 95% confidence interval is 2.2 ± 9.7 ms.

g. Since this is a repeated measures design, each analysis requires its own error term. In fact, the error half-bar for Number of Alternatives × S–R Mapping was 12 ms. Why do you think this error half-bar is so much larger than the 3.5 ms for (e)?

h. Find the 95% confidence interval for the nonadditivity of (g); interpret.

i. One 2×2 design remains. Do you expect the error half-bar for this interaction to be larger, smaller, or about the same as the two already given?

j. Suppose nonadditive effects of Stimulus Quality and S–R Mapping had been found. Why would it still be possible to maintain that stimulus encoding and response selection are distinct and independent processes?

Two stimuli

	R = S	R = S+1
Intact	329.6	348.8
Noisy	357.6	379.8

Eight stimuli

	R = S	R = S+1
Intact	371.4	472.4
Noisy	424.2	526.4

Two & eight stimuli

	R = S	R = S + 1
Intact	350.5	410.5
Noisy	390.9	453.1

(Mean reaction times in ms, after Sternberg, 1969, 1998. See further Section 11.4.4 as well as related Exercises 11.14a, 11.14b, and 20.c.5. I wish to thank Saul Sternberg for making these data available.)

19ANS. a. Stimulus Quality will affect the time required to identify the stimulus, a stimulus encoding stage, which comes early in the stimulus–response sequence. On the other hand, S–R Mapping will affect the time for response selection, after the stimulus has been identified. It thus seems a reasonable hope that these two variables will affect successive independent stages. This makes the experiment worth doing.

b. Number of Alternatives should influence the initial stage because identifying one of two possible stimuli seems easier than identifying one of eight. Number of Alternatives may also influence the final stage because selecting one of two responses seems easier than selecting one of eight. Hence Number of Alternatives would presumably be nonadditive with each of the other two variables, and this would appear in the two corresponding interaction residuals.

c. A little trial and error shows that Stimulus Quality and S–R Mapping yield near-parallelism, as hoped in (a), whereas both other 2×2 graphs yield nonparallelism, as expected in (b). I find parallelism easier to interpret than nonparallelism, so I would plot either Stimulus Quality or S–R Mapping on the horizontal, making a separate Stimulus Quality × S–R Mapping for each Number of Alternatives.

d. The near-parallelism of the 2×2 graphs for Stimulus Quality × S–R Mapping suggest that these two variables have additive effects and hence that they affect successive independent stages in the stimulus–response sequence.

The graph for Number of Alternatives × S–R Mapping is rather nonparallel, which implies that Number of Alternatives affects the second stage. The graph for Number of Alternatives

× Stimulus Quality is somewhat nonparallel, which suggests that Number of Alternatives may also affect the first stage. With only two curves of two points each, visual inspection of means is not too informative, so confidence intervals are needed.

e. The double difference of (e) is $(350.5 - 390.9) - (410.5 - 453.1) = 2.2$ ms.

f. To get the 95% confidence interval for (e), multiply the error half-bar of 3.5 ms by $t*(4) = 2.78$ to get a half-width of 9.7 ms. For repeated measures designs, the main component of error is usually individual differences in subject–treatment interactions. The data suggest no real effect for (e), which in turn suggests that these individual differences are negligible—that all individuals are additive—the conclusion needed theoretically.

g. Number of Alternatives and S–R Mapping are nonadditive. The amount of nonadditivity may be expected to vary across individuals, and these individual differences go into the error term, unlike (f).

h. The deviation from parallelism for the Number of Alternatives × S–R Mapping design is 80.9 ms. Multiply the given error half-bar of 12 ms by $t*(4)$ to get the 95% confidence interval of 80.9 ± 33.4 ms.

i. The deviation from parallelism in the Stimulus Quality × Number of Alternatives design is 23.9 ms, measured in the same way as previously. For the two previous measures of nonparallelism, the error half-bars were given as 3.5 and 12 ms. Larger error half-bars should be expected with a larger real effect because a larger real effect is likely to elicit larger subject–treatment interactions. Since this two-way nonadditivity is between the other two, I would guess the same for the error half-bar, say, midway between at 7 ms. This would yield a 95% confidence interval of 23.9 ± 19.5 ms. Since 0 is outside of this interval, the result is statsig. (The actual error half-bar was 6 ms, making the confidence interval a little shorter.)

j. Success of the additive-factors method depends on the joint hypothesis that the stage representation is valid and that each chosen variable influences only one stage. Inappropriate choice of the two variables could of itself produce nonadditivity even though the stage representation was valid.

20. To your dismay, you find in writing the report of your experiment on belief integration that you need to show that individual differences are real. This need is unusual, as everyone takes individual differences for granted, but your theoretical interpretation hinges on this point.

    a. From the material on page 163, construct a conservative test for individual differences.

    b. If you had foreseen this need, how could you have designed the experiment to get an appropriate test?

20ANS. a. The proper error term for subjects indicated by the expected mean squares on page 163 is σ_ε. Instead use MS_{SA} as the error. It is too large since it includes the interaction term, but this term will ordinarily be relatively small and so give a conservative test, but one that should surely demonstrate real individual differences.

b. To get a valid error term, replicate as indicated in the last paragraph on this page. This within cell replication provides MS_ε.

21. (With thanks to G. Keppel.) Perception of the external world is critical for all organisms: To locate food; to avoid predators; to seek warmth; and so on. A major task of comparative psychology is to study the diverse information processing systems of different organisms.

Desert iguanas, like other reptiles, are thought to use odors to guide their actions. They extrude their tongue to pick up airborne molecules, which they convey to the olfactory Jacobson's organ in their mouth. Tongue extrusion makes a good response measure and was used in the following inquiry into how iguanas perceive the external world (Pedersen, 1988). This kind of experiment can enlarge our appreciation of evolutionary processes that have shaped organisms' tools for information perception and goal-seeking.

The stimulus manipulation was a pan of sand placed in the test cage in which a single iguana lived for the seven days of the experiment. Five sources of sand were used in a repeated measures design: (1) clean sand; (2) sand from the iguana's home cage; (3) sand from a different iguana home cage; (4) sand from the home cage of western whiptail lizards, which often forage in close proximity to iguanas, although with minimal interaction; and (5) sand from the home cage of kangaroo rats, whose burrows iguanas favor for their own homes and whose droppings they apparently consume. Tongue extrusions were counted from a videotape of the 30 min following placement of the test pan of sand, with one test session per day. Subjects were 10 iguanas collected in the desert near Twenty-Nine Palms, California.

a. Pedersen preceded the five test sessions with two preliminary sessions. The first preliminary session was just after each iguana had been transferred from its home cage to the test cage, where it lived for the duration of the experiment. The second was at the same time on the following day. Why not begin the experiment at once?

b. The five test conditions for each subject must be run in some order. How would you handle this?

c. Pedersen considered the home cage condition as a control and planned to compare it with each of the other four conditions. Discuss the relative merits of planning these four comparisons versus planning to make the overall Anova.

d. The four comparisons planned by Pedersen in (c) are not independent. In particular, a substantial sampling error in the home cage condition would throw off all four comparisons en bloc. How might this danger have been reduced?

e. Pedersen presented mean response for each sand condition as a vertical bar, together with the standard deviation of the mean of the 10 responses (error half-bar or standard error of the mean) as a line projecting upward from the bar. Why is this error bar not appropriate for this repeated measures design? What alternative measure of variability would you suggest?

f. One iguana made no tongue extrusions under any test condition. This iguana seems extreme as overall mean response ranged from 6.0 in the home cage sand condition through 9.4 in the whiptail lizard sand condition to 23.1 in the kangaroo rat sand condition. Discuss the pros and cons of excluding this iguana from the analysis. What decision would you make?

g. The comparison of kangaroo rat sand with home cage sand was comfortably statsig, but none of the other three comparisons tested by Pedersen came close. Suppose this condition had also been nonstatsig; in that case, would you consider the experiment publishable?

h. These 10 iguanas are obviously not a random sample from any larger population. Yet they presumably have some generality. Based on the given information and your own background knowledge, what two considerations do you think are most important for generalization?

i. Keppel (1991, p. 365) used these data to illustrate the potential of specific comparisons, or contrasts, saying "A rich variety of comparisons can be examined with this experiment . . . . of perhaps greater interest [than Pedersen's four comparisons with home cage as control] are comparisons between conditions representing different odors—for example, home cage versus other iguanas, other iguanas versus lizards and kangaroo rats, lizards versus kangaroo rats, and so on. Certain complex comparisons may also be interesting: clean sand versus all other conditions combined, home cage and other iguanas combined versus lizard and kangaroo rat combined, and so on."

Make the best argument you can that the main interest in this experiment is with the sizes of the effects, and that the significance tests suggested by Keppel have relatively little value.

NOTE: The raw data are included in the Instructor's Manual. It is instructive to do repeated measures with and without the nonresponsive iguana. I am indebted to Geoffrey Keppel, who secured the raw data and included them in his text, and to Joanne Pedersen for permission to use these data. I wish to apologize to Geoff Keppel for criticizing his emphasis on contrasts, as his text is at the top in integrating statistics into empirics.

20ANS. The purpose of this lengthy exercise is to help illuminate the role of statistics in empirical science. See also final paragraph.

a. Exploratory behavior should be anticipated when any animal is put in a novel environment. Pedersen allowed two days for this to dissipate. This was well-done as mean tongue extrusions were 46 on day 1 but only 12 on day 2, and with little daily change thereafter. The adaptation period thus reduced the error variability. Also, it ensured that the statistical assumptions were better satisfied.

b. I would present the five conditions in a different order to each iguana so that conditions were balanced over days. A Latin square design would be just the thing (Section 14.3). Pedersen used a random sequence of treatments for each subject, which averages out position effects in the long run. A small number of subjects constitutes a short run, however, for which I feel systematic balance is safer. The Latin square also allows position effects to be measured, not just balanced.

c. The overall Anova does not localize any effects, whereas the four planned comparisons would. Exploratory responses would presumably have been adapted down in the home cage living, so home cage sand provides an appropriate baseline comparison. These four planned comparisons will have greater power than the overall Anova, moreover, if they are congruent with the real effects in the data. Furthermore, they do not require ε adjustment as does overall Anova.

The risk with the four planned comparisons is that they may not be congruent with the real effects. In that case, the overall Anova would be preferable, even though post hoc tests would be needed to localize effects.

A decision to use planned comparisons usually requires prior information about likely outcomes in the given situation. With desert iguanas, you probably lack prior information to make a firm decision. In your own line of investigation, however, you would have prior information as Pedersen did in hers.

d. Run the home cage condition more than once to get a more stable mean.

e. The standard deviation of the mean for each separate condition consists largely of differences between iguanas. But this error bar is irrelevant; only within iguana comparisons are of interest in this repeated measures design. These require a different error term, one that takes the correlation across iguanas into account. The error bars in Pedersen's graph are too large for the questions at issue.

Each repeated measures test should ordinarily have its own error term; hence the same mean may have different confidence intervals, depending on which other mean it is compared with. Pedersen restricted attention to just four comparisons, however, and it would seem appropriate to report a confidence interval for each.

f. Pro: Since this iguana is totally nonresponsive, it conveys no information about differences between the test odors. Also, the assumptions of normality and equal variance would seem threatened by this extreme case.

Con. This nonresponsive iguana is part of the population so eliminating it limits the generality of any test of the remaining iguanas. Further, eliminating this iguana is bound to raise the question whether it is justified. Keeping it in keeps this trouble out.

Pro. But keeping this iguana in dilutes the effect of the sand conditions, possibly yielding a nonstatsig result that might have been statsig without it. In that case, a post hoc test would be needed, which is weaker than if the elimination had been made by a priori plan.

Minor dilemmas like this are common in empirical research. Either action has probable and possible shortcomings. Note that the perplexity will largely vanish if the experiment is replicated. Here again, replication cures all.

If your face a similar situation, you may be able to reduce your decision anxiety by running Anova on Pedersen's data with and without the unresponsive iguana.

g. Nonstatsig results might be taken to suggest that iguanas make little use of odor in conditions analogous to those of their desert life. But this conclusion seems unattractive because odors are common biological informers in other reptiles, as stated, not to mention insects and mammals (e.g., bloodhounds). Moreover, the tongue extrusion response and the Jacobson's organ for odor reception would seem evolved for just this purpose. Nonstatsig results would thus suggest some shortcoming in experimental procedure, so the results would not be publishable.

Note in light of this argument how the kangaroo rat sand condition validates the experimental procedure.

h. Generalization of these iguana results must rely on extrastatistical inference. Two interrelated considerations seem important. One is the results themselves, which demonstrate discriminative capability of the tongue reaction; the other is biological, namely, that odor perception is a general and useful sensory process that requires specific physiological mechanisms. On this basis, I would expect these results to hold for most iguanas throughout the Americas. Open to question, however, is the extent to which the specific response levels in this experiment depend on previous experience of iguanas in their desert environment.

i. In this experiment, as in many others, I feel the main concern is to estimate the size of the effect in each condition and that significance tests have limited relevance. Of course, a significance test is needed to show that the data represent some real effect over and above error variability, as noted in (g). And confidence intervals are in order as part of the estimation of effect size. Beyond that, however, I feel the results should be taken as estimates of the effect sizes, without great regard for statistical significance.

The only yes-no decision required with these data is whether to publish. Granted that they are publishable, they seem best appreciated as suggestions about where to go next. Indeed, the relatively low response of 7.4 to other iguana sand has substantive interest that contributes to understanding of how sensory processes have evolved and how they adapt to local conditions.

In this view, the study should be considered information about sizes of effects. For example, the mean response of 9.4 to whiptail lizards was not statsig larger than the mean response of 6.0 to the home cage. Nevertheless, from the cited information given by Pedersen about daily life of iguanas, this difference may well be real even though small compared to the response of 23.1 to kangaroo rats. No critical decision rests on the reaction to whiptail lizards. The given data provide a proper estimate of this effect size. This kind of assessment can provide helpful clues for follow-up work.

The numerous significance tests suggested by Keppel illustrate the flexibility of contrasts. I feel, however, that they obscure the main value of these particular data.

TONGUE EXTRUSIONS BY DESERT IGUANAS

Subject	Clean sand	Home cage	Other iguanas	Whiptail lizards	Kangaroo rats
1	24	15	41	30	50
2	6	6	0	6	13
3	4	0	5	4	9
4	11	9	10	14	18
5	0	0	0	0	0
6	8	15	10	15	38
7	8	5	2	6	15
8	0	0	0	11	54
9	0	3	2	1	11
10	7	7	4	7	23

NOTE: After Pedersen (1988) and Keppel (1991).

ANSWERS FOR CHAPTER 7

1. Plot the following two data tables for a $3 \times 3 \times 2$ design.

   a. By visual inspection, say which main effects are/are not substantial.

   b. By visual inspection, say whether the AB residual is/is not substantial.

   c. Do (b) for the ABC residual.

   d. Make rough visual estimates of the marginal row and column means, and plot each as two curves in a 2×3, CA or CB design. By visual inspection of these graphs, say whether these residuals are substantial or not.

	C_1		
	B_1	B_2	B_3
A_1	7	21	39
A_2	14	23	30
A_3	23	24	25

	C_2		
	B_1	B_2	B_3
A_1	20	32	48
A_2	26	36	41
A_3	32	34	36

1ANS. a. Substantial main effect of B, shown by the strong upward slopes at two levels of A. Substantial main effect of C, shown by the difference in overall elevation for the two 3×3 graphs. Slight main effect of A, as may be seen by making rough visual estimates of the marginal row means in each 3×3 design. In this example, the three row means in each 3×3 design roughly equal the values in the middle column, and these are roughly constant. (Although A has substantial effects at both B_1 and B_3, these two effects average out in the crossover pattern at each level of C.)

b. Substantial AB residual, as shown by the similar pattern of nonparallelism in both 3×3 graphs. (Opposite patterns of non-parallelism in the two separate 3×3 graphs could cancel out, leaving little AB residual. For a real case, see Note 7.5.1b.)

c. Slight ABC residual by the pattern similarity of the two 3×3 designs.

d. Visual inspection of these two graphs shows that the two curves are roughly parallel, indicating slight AC and AB residuals.

2. In Table 7.2:

   a. Get the three systematic SSs for the factorial Anova, assuming $n = 1$.

   b. Assuming the observed pattern mirrors the true pattern, in what two ways is the factorial Anova bungling the data analysis?

2ANS. a. All three systematic sources have SS = MS = 4.

b. The two-way factorial Anova is bungling the data analysis in two ways. First, it is fractionating the one real effect into three parts, none of which properly characterizes the pattern. Second,

it is thereby losing power. In short, factorial Anova is at once less informative and less powerful.

But it is not Anova that is bungling; it is the investigator who applies Anova from habit, thinking that statistics can replace understanding.

3. Suppose you had expected the blame judgments of Figure 7.1 to come out as they did. How would you plan to analyze the data?

3ANS. To specify the pattern in Figure 7.1 needs two tests. First, a 2×4 Anova of the Malice and Displacement curves. This should show both main effects statsig, but not the interaction. Second, a one-way Anova on the Accident curve. This should show no main effect. (A test of the linear trend of this curve would have greater expected power.)

4. Use computer Anova to show that SS_{AB} has the same value for all three of the 2×2 graphs in Figure 7.2. Comment.

5. In Exercise 27 of Chapter 5, suppose the 6 in cell 24 is changed to deviate from the parallelism pattern in the same way as the entry of 3 in cell 34. Find the interaction residuals in column 4. Comment.

5ANS. Since the entry of 3 in cell 34 deviates from parallelism by being 6 too low, the same deviation from parallelism of the 6 in cell 24 requires it be 0. Following Exercise 5.27, the interaction residuals in column 4 are then $3.00, -1.50, -1.50$, reading down. The largest residual is thus in the cell that doesn't deviate from the overall parallelism pattern. The moral is that residuals, contrary to common statistical advice, may not be too helpful. This is no surprise after the lesson of Exercise 5.27, in which a single deviant point caused a residual to appear in every cell of the design. Once again, the arbitrary Anova model yields a clumsy analysis.

The next five exercises consider empirical situations that involve the two basic issues in understanding interactions: *model* and *measurement*. Some require background information that makes them longer.

6. In the experiment of Table 7.1, P and Q are disturbed by their disagreement about the interaction. Their first step is to check for mistakes in the recorded data. Sure enough, the time of 6 s is really 5 s.

   a. Does this change affect the direction of P's interaction?

   b. Does this change introduce an interaction in Q's data?

   c. Psychologically, what does each interaction mean in this experiment, taking each at face value?

   d. What is the moral of this exercise?

6ANS. a. P's interaction has the same direction with 5 as it had with 6.

b. With this correction, Q's data are not parallel; they exhibit an Anova interaction.

c. Psychologically, the two interactions disagree. P's interaction says that reward has more effect at low motivation; Q's says that reward has less effect at low motivation.

d. The moral of this exercise is that not merely the size, but even the direction, of an interaction depends on the measurement scale.

7. Apply the equation for the proportional change model given in Note 7.3.2b to verify the arithmetic given with Figure 7.3.

8. This exercise refers to the Rosenthal–Rosnow data in Note 7.3.5a. From Equation 5.2a, the expression for the interaction residual in any cell of a two-way design is

[cell mean − row mean − column mean + overall mean].

   a. Use this expression to show that the interaction residuals in the Rosenthal–Rosnow data are +1 or −1.

   b.* Show that any 2 × 2 table with nonparallel data will yield a pure crossover for the factorial graph of interaction residuals. How does this relate to the interaction df?

8ANS. a. By the given expression, the residual in cell 11 is

$$\text{residual}_{11} = 2 - 3.5 - 2 + 2.5 = -1.$$

The other three residuals can be calculated similarly. The factorial graph of these residuals exhibits a pure crossover. This crossover is the basis for claiming that "conservatives and liberals reacted in exactly opposite ways to the two types of propaganda."

b.* To prove (b), you could transcribe the above numerical pattern into algebraic form, using the Anova model of Equation 5.2a. However, the following verbal reasoning seems more illuminating. The residuals are what are left in the cells after subtracting out the overall mean, the row effect, and the column effect. Therefore, if the residual is inserted in each cell of the factorial design, the row means, the column means, and the overall mean of these residuals must be zero. But the Anova interaction has only one df in a 2 × 2 design; if we know one residual, we know all four. To make the marginal means zero, the residuals must be equal and opposite in each row and in each column.

9. The multiplication model,   Behavior = Expectancy × Value, has been conjectured in diverse fields in psychology. But valid tests have been elusive. Potential for model test with factorial Anova is indicated in this exercise.

   a. Assume numerical values of four levels of Value, two positive and two negative, and of three levels of Expectancy. Plot the factorial graph, with Value on the horizontal. How does the shape of this graph reflect the multiplication operation?

   b. Suppose that an outcome with negative Value causes subjects to change their Expectation. How will this interaction appear in the factorial graph?

9ANS. a. One useful set of levels is {−3, −1, 1, 3} for Value and {.2, .5, .8} for Probability. With Value on the horizontal, the factorial graph will be a two-wing fan of three curves intersecting at zero Value. The multiplication model says each curve is a straight line, with slope equal to the Probability. Statistically, all the interaction SS should be concentrated in the row:linear × column:linear trend.

   In fact, attempts to test multiplication models ran into trouble for quite some time because no one knew how to measure the subjective psychological values of the column stimuli, which are

essential to the test. This difficulty was overcome with the linear fan theorem of functional measurement theory (Section 21.5).

b. If subjects change their Probability when Value is negative, the left wing of the fan will change slope, since Probability equals slope.

10. (After Bogartz, 1976.) To test whether younger children are more distractible, Hale and Stevenson (1974) presented 5- and 8-year-olds with a task of short-term visual memory, both with and without distractors. This is a 2 × 2 design, with two levels of age and two of distraction.

For the 8-year-olds, mean scores were 6.43 and 5.65 without and with distraction. Corresponding scores for the 5-year-olds were 3.94 and 2.88. (Perfect score was 12.)

The problem is to compare the distraction effect between age groups; this comparison is complicated by the large main effect of age. Hale and Stevenson compared the two differences: (6.43 − 5.65) = .78 and (3.94 − 2.88) = 1.06. The difference between these two differences is the Age × Distraction interaction, which was not statsig. Since this experiment had used 72 children at each age, the authors considered they had adequate power to conclude that both age groups were equally distractible. Their design seemed an ingenious way of factoring out the large main effect of age on memory.

But Bogartz (1976) pointed out that their interpretation rested on an implicit—and dubious—theoretical assumption, namely, that the distraction process obeyed an additive model. Suppose instead that the proportional change model applies, so that the decrement due to distraction is proportional to the level of memory under no distraction. Then equal proportional distraction would yield nonzero Anova interaction and zero Anova interaction would correspond to unequal distraction proportions.

   a. Why was it important to have a substantial main effect of distraction?

   b. Why does the main effect of age have secondary interest in the theoretical analysis? Exactly what interest does it have?

   c. Assume with Bogartz that the distraction effect follows a proportional change model. Let w denote the proportion by which memory decreases from the nondistraction to the distraction condition. If $w = .20$, show that this proportional change model implies a nonzero Anova interaction (nonparallelism) when this distraction proportion is equal across age.

   d. Generalize (c) to allow any value of w.

   e. Construct a numerical example to show that zero Anova interaction implies greater distractibility for 8-year-olds under the proportional change model. Show this result is general, not peculiar to your numerical example.

   f. Give a tentative interpretation of the cited data in the light of the foregoing considerations.

   g. (Optional exercise; not in text). Bogartz' model took account of a complication that the response was multiple choice, so the children could be correct by guessing. Complete distraction thus corresponded to a score that may be taken as 2.85 (Anderson & Well, 1975), not 0 as assumed above. Redo the data analysis to assess whether both age groups were equally distractible under the proportional change model.

10ANS. a. It was important that distraction have a substantial effect because the purpose was to compare the size of this effect across age. If the effect itself was small, any difference in its size across age would be even smaller, so the experiment would lack power.

Previous work was thus necessary to develop an experimental task not too hard and not too easy. Other than this, the distraction main effect had no particular interest.

b. Older children did better, of course, but this has minor theoretical importance. It does, however, complicate the comparison of distractibility across age because some way must be found to "adjust" for the difference in performance of the two age groups.

c. Suppose distraction reduces recall by 20%, the same for both age groups. Under the proportional change model, the predicted distraction effect is $.2 \times 6.43 = 1.29$ for 8-year-olds, and $.2 \times 3.94 = .79$ for 5-year-olds. The difference, $1.29 - .79 = .50$ is the numerical value of the Anova interaction. In principle, this interaction is real.

d. Let Y_5 and Y_8 denote the memory scores for the two age groups on the nondistraction test. Let w denote the proportionality coefficient for decrement; this is equal for both ages by assumption. Then the amount of decrement is $w Y_5$ and $w Y_8$ for the age groups. The difference between these two decrements is the Anova interaction: $w(Y_8 - Y_5)$. This difference will be nonzero as long as the two nondistraction memory levels differ, as is expected from the age effect.

e. Suppose distraction decreases the memory score by 1 point, the same for both age groups. This corresponds to parallelism, or zero Anova interaction in the factorial graph. Under the proportional change model, however, the distraction effect is $1/6.43 = 16\%$ for 8-year-olds and $1/3.94 = 25\%$ for 5-year-olds, which are unequal. Because the denominators differ, the inequality holds generally, for any nonzero number in place of 1.

f. With a chance level of 2.85, w estimates are .97 and .22 for 5- and 8-year-olds, respectively. Although statistical analysis is needed, the younger children do seem more distractible.

11. This exercise illustrates a not uncommon situation in which an interaction is claimed, but the evidence falls critically short.

An interesting field study by Ellsworth and Langer (1976) sought to show that the effect of staring depended on the social situation. Specifically, their hypothesis was that staring "may function as a stimulus either to approach or to avoidance depending on the context" (p. 117). (All quotes from their article.)

Their unwitting subjects were female shoppers just inside or outside a large department store, who had no idea they were in an experiment. Each shopper was told by one experimental assistant, A, that another experimental assistant, B, in view close by, needed help. Two variables were manipulated in a 2×2 design. **Ambiguity:** A indicated to the shopper either a *clear* condition (saying that B had lost her contact lens and needed help) or an *ambiguous* condition (saying that B was not feeling well and needed help). **Staring:** B either stared at the shopper or looked at the ground. Of main interest were ratings on a 0–4 scale by a blind observer of how much the subject helped B. For example, a shopper subject "who appeared to hesitate but went

no further" was rated 1.

a. The main hypothesis was that "the difference in the amount of assistance offered to the two types of victims [clear or ambiguous] was expected to be greater when the victim stared" (p. 118). "The hypothesis was confirmed: There was no main effect of staring. More help was elicited in the clear condition than in the ambiguous, but the difference between these two conditions only reached significance when the victim [B] stared" (p. 117). Does this result provide adequate evidence for the conclusion?

b. The authors also state that "The Stare $\times$ Ambiguity interaction was nonsignificant, $F(1, 44) = 1.76$" (p. 120). How does this bear on the conclusion quoted in (a)?

11ANS. The main hypothesis was that the difference in amount of assistance offered to the two types of victims would be greater when the victim stared. This hypothesis is exactly equivalent to an interaction in the 2×2 design. The key issue is to demonstrate this interaction.

The evidence quoted from page 117 fails to confirm the main hypothesis. The main hypothesis implies the interaction will be statsig, but it was not. The fact that one difference is statsig and the other not does not imply a real difference between the two differences. This fallacy is by no means uncommon.

The quoted claim obscures this failure by citing a statsig effect in the stare condition and a nonstatsig effect in the no-stare condition. This evidence is inconclusive, of course, for the confidence interval for the difference between the two means may well contain 0.

b. The nonsignificance of the interaction contradicts the quoted conclusion and infirms the main hypothesis.

This example also illustrates how confusing it can be to unravel the claims of an article to assess the evidence on which these claims rest. It is thus pertinent to note that this article was criticized on the ground that it had too little power to warrant acceptance of the null hypothesis of no main effect of staring by Oakes (1986, p. 11), who failed to realize that this test was beside the point.

12. In the upper left quadrant of Figure 7.4, left and right panels have the same shape. Suppose the upper left point is displaced upwards one unit, from 3 to 4.

a. Express the three-way interaction as a triple difference to show that the one-unit change makes it nonzero.

b. Show that the three-way interaction can be made zero again by a one-unit change in any of the eight data points.

c. What does this imply about interpreting observed interactions?

12ANS. a. With the one-unit change in bold face, the triple difference for the three-way interaction is

$$[(4 - 0) - (4 - 3)] - [(5 - 2) - (6 - 5)] = 1.$$

b. Visual inspection of the equation just written shows that subtracting 1 from any positive entry or adding 1 to any negative entry yields a sum of 0.

c. This example illustrates the difficulty of interpreting a statsig interaction in such a small design. Even if it were known to be localized in a single point, that single point could be any one of the eight.

If the direction of interaction is predicted, of course, interpretation is often simplified. With larger designs, moreover, extrastatistical information about expected trend can help localize an interaction.

13. You do a 2 × 2 experiment, planning a regular factorial Anova. Unexpectedly, the entries in the first row are 2 and 1, those in the second row 1 and 9.

    a. Why does the factorial Anova seem inappropriate for this data pattern?

    b. Suggest an alternative analysis and discuss pros and cons.

13ANS. a. At face value, the data suggest three equal means, one different. This is the pattern discussed for Exercise 2.

b. First, I would run the 2 × 2 Anova as planned. By Exercise 2, the data pattern implies that all three sources will tend to be statsig or nonstatsig together. If the 2 × 2 Anova shows statsig results, visual inspection shows that the pattern of data is similar to that of Table 7.2, suggesting just one real difference. The firmness of this conclusion of just one real difference would depend on the narrowness of the confidence interval.

If the 2 × 2 Anova is not statsig, post hoc application of the two tests cited in the text for Table 7.2 would be appropriate. Post hoc tests are somewhat untrustworthy, of course, the more so in this case because they disagree with prior expectation.

Second, I would judge whether this pattern of results, if real, represents worthwhile knowledge. As one possibility, some configural effect may be involved that requires both factors to be above some threshold. Such judgment would rest on extrastatistical considerations related to the particular experimental situation. If it looks important enough to warrant public attention, replication seems more or less essential unless perhaps the confidence intervals were very narrow. If it did not look important, I would curse my data and push on to some new thing.

14. Suppose the four levels of B are 1, 2, 3, and 4; that A has one constant effect if $B < 2.5$, another constant effect if $B \geq 2.5$; and that the response equals the sum of the values of A and B. For simplicity, assume zero error variability.

    a. How will this process appear in a factorial graph?

    b. In what way would Anova misrepresent the specified process?

14ANS. a. The factorial graph consists of two line segments, one for the two B levels below 2.5, the other for the two B levels above 2.5. These two segments would be misaligned unless the two constants are equal.

b. Anova reports no interaction, whereas the data result from an interactive process, for the effect of A does depend on the level of B.

15. In a Person × Situation design, suppose persons and situations are perfectly additive—except for one deviant person.

    a. Extrapolating from Exercise 27 of Chapter 5, say how Anova will misrepresent these data.

    b. How could visual inspection reveal the truth?

15ANS. a. Extrapolating from the cited exercise, the one deviant person will cause an interaction residual to appear in every cell of the Person × Situation design. An actual case from a different area in which a single deviant point caused a statsig interaction is in Figure 5.3 (page 131).

b. In principle, visual inspection will show near-parallelism for all but one of the Person curves in the design, as illustrated in Figure 5.3.

The use of standard Anova to study Person–Situation interaction was a desperation tactic reflecting the lack of adequate tools for a difficult, important problem. *Personal Design* (Chapter 11) can provide some leverage.

16. In a 2^3 design, construct hypothetical data with all effects zero except:

    a. A.   b. A and B.   c. AB residual.

    d. ABC residual.   e.* AB, AC, and BC residuals.

Note: Simplify by using cell entries of $+1$ and -1 except use also 0 for (b).

(Trial and error will yield plenty error on (e); use Equation 13 of Chapter 5.)

16ANS. c. Make a 2 × 2 design with $+1$ and -1 in the first row, -1 and $+1$ in the second row. This pure crossover yields only the AB residual. Use for both levels of C.

e*. Start with the 2^3 design of (c); this represents $(\alpha\beta)_{jk}$ in Equation 5.13. Make a similar 2^3 design that has only the AC residual, the $(\alpha\gamma)_{jm}$ in Equation 5.13, and another for the BC residual, the $(\beta\gamma)_{km}$. Add the entries cell by cell. This yields 3 in cells $A_1B_1C_1$ and $A_2B_2C_2$, and -1 in the other 6 cells. An easy check is to give the data to the computer. Only the three indicated SSs should be nonzero.

17. What is your psychological interpretation of the configural effect shown in Figure 7.1?

ANSWERS FOR CHAPTER 9

1. a. Use the regression equation given for Figure 9.1 to predict Y for new cases, with X_{new} = 0, 2, 5, and 12.

   b. Which of these predictions is most/least reliable?

   c. Will all predictions for new cases lie on the regression line?

1ANS. a. The regression equation is $Y = 2 + \frac{1}{2}X$. This is the line of best fit to the given data and hence is used for prediction for new cases. Substitution of the given X_{new} yields Y_{new} = 2, 3, 4.5, and 8 for the four respective cases.

b. Following the subsection on *Confidence Bands*, the most reliable prediction is for X_{new} = 5, which is at the center of the X range. The least reliable is for X_{new} = 12, which is farthest from the center of the X range.

2. Justify the statement: "In Figure 9.1, visual inspection indicates that the regression line should pass through the Y values of 3, 4, 5, and 6 for the four successive values of X."

2ANS. Visually, this line comes equally close to all four data points; all deviations equal 0.2. Another line could come closer to two of the data points, but it would be farther from the other two. Hence it would yield a larger sum of squared deviations.

3. a. Use hand calculation to show that the standard deviation, or error half-bar, for b_1 for the data of Table 9.1 is .063. Use calculations given in the text.

   b. Find the 95% confidence interval for b_1.

3ANS. a. The standard deviation of b_1 is $\sqrt{MS_{dev}}/\sqrt{SS_X}$. Table 9.2 lists MS_{dev} = .08; the text lists SS_X = 20. Thus, the standard deviation for b_1 is $\sqrt{.08/20}$ = .063.

b. On 2 df, t^* = 4.30; .063 × 4.30 = .27. The 95% confidence interval is .5 ± .27.

(Although the N of 4 is unrealistically small, this example gives the essential idea.)

4. a. Graph these data and find the linear regression by visual inspection.

$$X = 2, \ Y = 5.8$$
$$X = 4, \ Y = 5.2$$
$$X = 6, \ Y = 4.2$$
$$X = 8, \ Y = 2.8.$$

   b. How is this example related to Figure 9.1 in the text?

4ANS. a. $b_1 = -\frac{1}{2}$, $b_0 = 7$. The rationale for this line is the same as in Exercise 2.

b. These data are order-inverted from Figure 9.1, so b_1 is equal and opposite.

5. Under *Unequal Variance* in Section 9.1.4, verify that the Y values of 3 ± 10 would yield b_1 values of −4 and 6.

5ANS. This is straightforward calculation, easy enough by hand.

6. Graph $Y = b_0 + b_1X$ over the range from −10 to 10 for:
   a. $b_0 = 0$; $b_1 = -2, -1, 0, 1$, and 2.
   b. $b_0 = -2, -1, 0, 1$, and 2; $b_1 = 1$.
   c. Explain the pattern in (a) and in (b).

6ANS. A good graph will include all five lines together to make the pattern clear. Each line of course requires only two points for graphing. The pattern in (a) is a two-wing linear fan, reflecting the varying slope, with all curves intersecting at the origin. The pattern in (b) is a set of parallel curves, with elevations given by the values of b_0.

7. If Galton had studied the relation between heights of mothers and daughters, what do you guess he would have found?

7ANS. My guess is that b_1 should be nearly the same, perhaps a little higher, for mother–daughter as for father–son. Empirical support for this guess comes from Pearson and Lee (1903), who found slope coefficients near ½ for height relations of father–son, father–daughter, mother–son, and mother–daughter, each based on upwards of 1000 cases. Thus,

$$H_{daughter} = \overline{H}_{mother} + \frac{1}{2}(H_{mother} - \overline{H}_{mother}),$$

where $\overline{H}_{mother}$ is the mean height of all mothers in the population.

Also, Galton would definitely have found regression to the mean: Daughters' heights are less extreme, on average, than their mothers' heights.

8. Do-it-yourself example of subgrouping artifact in correlation. (No calculation is needed or wanted; a graph will suffice.)

   a. Consider two subgroups of three cases each. Make up simple artificial data with zero correlation within each subgroup, but a positive correlation for the group as a whole.

   b. Show similarly how a substantial positive correlation within each of two subgroups could vanish in the group as a whole.

8ANS. a. Subgroup 1: $X = 1, 2, 3$; $Y = 1, 1, 1$.
      Subgroup 2: $X = 2, 3, 4$; $Y = 2, 2, 2$.

Each separate subgroup shows a slope of 0. In contrast, the set of all 6 data points shows a slope of 1.

b. Subgroup 1: $X = 1, 2, 3$; $Y = 1, 2, 3$.
   Subgroup 2: $X = 4, 5, 6$; $Y = 3, 2, 1$.

Each separate subgroup shows a slope of ± 1. In contrast, the set of all 6 data points shows a slope of 0.

9. An article reports a correlation of .31 for Y and X. In terms of variance, how much predictive power does X have? What does this mean in practical terms?

9ANS. Less than 10%, since $.31^2 = .0961$.

10. One point can make a big difference in regression. Run two regressions for the following data, one for the first 8 data points, the other for all 9, and compare them. What is the moral?

X	1	3	5	6	7	8	10	12	18
Y		10	11	10	12	12	10	11	10	19

10ANS. With the first 8 data points, $b_1 = 0$. With the one added point, the regression yields $b_1 = .41$, $F = 7.56$, statsig on 1/7 df.

The difference between the two regressions comes partly from the increased ranges of X and Y, partly from the apparent change in the $Y-X$ relation over the larger range. Moral: Always scrutinize the scatter plot. And be cautious about extrapolating a regression equation outside the given range of X.

11. In a psychophysiological study of emotion, investigator R measured X and Y for each of 6 subjects on each of 2 days. He reported a 95% confidence interval for b_1 of 2 ± 1.86 on 10 df. He argued that this statsig $Y-X$ relation verified his own theory of emotion and infirmed the James–Lange theory of emotion (that the conscious emotion is not the cause of the emotional reaction, such as fight, flight, or freeze, but a consequence). Why does R have only 5 df? What is his actual confidence interval?

11ANS. Since R reported 10 df, he evidently ran the regression using all $2 \times 6 = 12$ data points. But this analysis violates independence since the two measures for each subject are correlated.

A correct analysis would use one score for each subject, namely, the mean of the two measured scores. This will give the same slope, but with only 4 df. Accordingly, R's confidence interval used too small a value of t^*. It needs to be lengthened by the ratio of t^* for 4 and 10 df: $2.78/2.23 = 1.25$, This yields the confidence interval of 2 ± 2.32, which is not statsig. For R personally, however, it might be close enough to warrant more work.

This example illustrates *double counting*, which is more common and more serious with frequency data (Chapter 10). Keep a weather eye out for double counting when you read a published report. Scrutinize the df in published articles; they can reveal inappropriate regression analyses just as well as inappropriate analyses of factorial design. (For a published example of double counting with Anova, see Anderson, 1982, p. 295.)

12. The speed of a ball rolling to the bottom of Galileo's inclined plane is proportional to the square root of the height from which it started to roll. You wish to test whether human cognition follows this physical law. In your experimental situation, you show subjects a ball at various elevations on an inclined plane; they predict how fast it would be rolling at the foot of the incline if it was released to roll freely down. Subjects respond on a graphic rating scale. In the instruction period, a few actual ball rolls are shown to calibrate their response scale. Each subject receives three trials at each of six elevations (Karpp & Anderson, 1997).

  a. Why would it be obtuse to test the null hypothesis that the initial height of the ball affects subjects' responses?

  b. Why would it be a mistake to fit a regression line to the

data for all the subjects together?

  c.* For plotting the data, why does Galileo's law of falling bodies suggest an advantage from using $\sqrt{\text{height}}$ on the horizontal axis instead of height?

12ANS. a. It should be obvious to nearly every subject that the higher the ball, the greater will be its speed at the bottom of the inclined plane. To apply a significance test that is totally unnecessary shows a fearful mind, enslaved to statistics.

b. The data are not independent since repeated measures were made on each subject. A valid error term or confidence interval could not be obtained.

c. With $\sqrt{\text{height}}$ on the horizontal, the physically correct values of the data will plot as a straight line; deviations from linearity will be easy to detect by visual inspection. With this in mind, it may be desirable to use levels of the height variable that are more equally spaced on the square root scale than on the height scale itself.

The alternative of using height on the horizontal may seem more natural. But then the physical law appears as a curved line, deviations wherefrom would be hard to judge.

13. What causes your feelings of hunger? Are there hunger receptors like receptors for vision and taste? If so, where are they?

Cannon (1934) hypothesized that hunger is caused by stomach contractions. His devoted student, Washburn, trained himself to swallow an uninflated rubber balloon with a rubber tube that led out through his esophagus and mouth. Cannon then inflated the balloon inside Washburn's stomach. Stomach contractions would increase air pressure in the balloon, which were recorded on an apparatus not visible to Washburn. Sure enough, when Washburn reported hunger pangs, the apparatus recorded stomach contractions. Cannon concluded that stomach contractions caused the feeling of hunger.

  a. Washburn's stomach supported Cannon's hypothesis in this classic experiment. What is the main objection?

  b.* What alternative approaches can you suggest as further tests of Cannon's hypothesis?

13ANS. a. The joint occurrence of stomach contractions and hunger pangs does support Cannon's hypothesis that the former caused the latter—but not too strongly. This is only a correlation, lacking experimental control.

  b.* Injecting sugar into the blood will stop the stomach contractions but not the hunger. Furthermore, humans with stomachs surgically removed still experience hunger pangs.

Despite the eventual fate of this particular claim, Cannon has high scientific renown and was cited by Burke (1998) in his regular column, "Connections," in *Scientific American* as the only psychologist to have a mountain named after him—Mt. Cannon in Glacier National Park, in recognition of his pioneer ascent of this rugged 10,000' peak together with his wife, Cornelia, on their honeymoon in 1901; see Cannon, 1945, pp. 23*ff*. They left a note in a bottle under a cairn found in 1985 (see Benison, Barger, & Wolfe, 1987, p. 422, n. 43.)

Cannon's study of hunger has been a classic in introductory psychology texts, but contrary to Burke, Cannon was a physiologist, not a psychologist. However, there is a psychologist of renown, Richard C. Atkinson, previously chancellor at UCSD and currently president of all the campuses of the University of California, who has a mountain named after him, Mt. Atkinson in Antarctica, from his term as Director of the National Science Foundation.

14. Table 9.3 lists four sets of data ingeniously constructed by Anscombe (1973). Run a computer regression and get plots for each set.

TABLE 9.3

FOUR SETS OF REGRESSION DATA

Case	Y_a	Y_b	Y_c	X_{abc}	Y_d	X_d
1	8.04	9.14	7.46	10.0	6.58	8.0
2	6.95	8.14	6.77	8.0	5.76	8.0
3	7.58	8.74	12.74	13.0	7.71	8.0
4	8.81	8.77	7.11	9.0	8.84	8.0
5	8.33	9.26	7.81	11.0	8.47	8.0
6	9.96	8.10	8.84	14.0	7.04	8.0
7	7.24	6.13	6.08	6.0	5.25	8.0
8	4.26	3.10	5.39	4.0	12.50	19.0
9	10.84	9.13	8.15	12.0	5.56	8.0
10	4.82	7.26	6.42	7.0	7.91	8.0
11	5.68	4.74	5.73	5.0	6.89	8.0

NOTE: First three sets of Y all have same X values, listed in the fourth data column. From Anscombe (1973).

a. What is striking about this example?

b. What is the moral of this example?

14ANS. a. In this remarkable example, all four sets of data yield identical values of b_0, b_1, and r. Without looking at the actual data, one could not realize that each set shows an entirely different relationship.

b. Moral: Plot your data and scrutinize the plot.

15. In terms of the variance of Y, justify the statement under Equation 1 that "A horizontal regression line means that X has no predictive power."

15ANS. A horizontal regression line means $b_1 = 0$, so the predicted $\hat{Y}_i = \bar{Y}$, for all i. Hence $SS_{deviations} = \sum (Y_i - \hat{Y}_i)^2 = \sum (Y_i - \bar{Y})^2 = SS_Y$. Hence the regression line does not decrease the variance of Y.

This may also be seen indirectly by noting that $b_1 = 0$ implies $SS_{pred} = 0$ by Equation 7b, and hence again that $SS_{dev} = SS_Y$ by Equation 8b.

It would not be correct to say that the regression line has no predictive power; $\bar{Y}$ is a better predictor than any other value. What has no predictive power is X.

16. A single group of subjects is run, each subject serving at four levels of a metric variable A. A regression line is fitted to each subject's data to obtain a slope value b_1 for each subject.

a. You are told to apply Anova to these slope scores. What do you do?

b. Suppose three independent groups of subjects were run, each at one level of variable B. How would the Anova change? How would it be interpreted?

c. Compare this analysis with the alternative of a repeated measures Anova of the $(S \times A) \times B$ design.

16ANS. a. Treat the b_1 scores as Y scores in an Anova. This yields $F_{mean} = MS_{slope}/MS_{error}$, on 1 and $N - 1$ df. F_{mean} tests H_0: $\mu_{slope} = 0$.

b. In this three-group design, F_{mean} tests the null hypothesis of zero mean slope, averaged across levels of B. F_B tests the null hypothesis of equal true slope between the three B conditions.

c. The slope coefficients represent the linear trend of A, which is usually more sensitive than the A main effect in the Anova. This principle was illustrated with the numerical trend example of Chapter 4. If there was some interest in nonlinear trend, it could be assessed similarly using nonlinear trend coefficients (Section 18.2) or with multiple regression (Chapter 16).

17. For repeated measures regression, how would you find the confidence interval for the mean b_1, averaged across subjects?

17ANS. Treat the b_1 scores, one for each subject, exactly as the Y scores in Chapter 3.

18. Suppose the regression line has slope of 0. Then it says to predict $Y_{new} = \bar{Y}$, regardless of X. What property of the mean makes this a best prediction?

18ANS. The mean comes closest to all the individual values in the sample because the deviations from the mean, $Y_i - \bar{Y}$, always sum to zero.

(But because the sample mean is an interval of uncertainty, its variance must also be considered. For a normal distribution, the sample mean has the lowest variance of all measures of central tendency. For some nonnormal distributions, however, other measures of central tendency, such as the median, may have lower sampling variability.)

19. Below Equation 2, the text states "Hence a positive value of $(Y_i - \bar{Y})$ will tend to be paired with a positive value of $(X_i - \bar{X})$." Does this work in the case in which all X_i are negative? (Try a numerical example with three data points.)

19ANS. Yes; "positive value of $(X_i - \bar{X})$" refers to a *deviation* from $\bar{X}$. The larger values of X_i are those that are less negative than $\bar{X}$. These will yield positive *deviations*, which will tend to be paired with the positive deviations for Y."

20. Under *Unequal Variance* in Section 9.1.4, show that variances of 0, 100, 0 for $X = 1, 2, 3$ produce zero variance in b_1. What practical implication follows from the contrast between this case and that cited in the text?

20ANS. The value of Y at the middle X changes the elevation of the regression line, but by symmetry can have no effect on the slope. The obtained slope will thus always be 1. This symmetry operates in Equation 2 for SP_{YX}, in which $(X_i - \overline{X}) = 0$ for $X = 2$. This term in the sum is thus zero, regardless of the value of Y_i.

The practical implications derive from the difference between the minimal effect of a deviation near the middle of the X range and the maximal effect of a deviation near the end of the X range, as in the text example. Extreme scores are more dangerous near the ends of the range, where they are often more likely to boot.

21. Let $Y = 2, 1, 6$ for $X = 1, 2, 3$. Graph the data and use visual inspection to find the best-fit linear regression using least absolute deviations.

21ANS. The sum of absolute deviations is minimized for the line that joins the first and third points. (The text notes that a best-fit line must pass through two data points.)

22. Show that Galton's equation implies regression to the mean with (a) a numerical example and (b) algebraically.

22ANS. Regression to the mean follows because the term in parentheses is an adjustment that depends on the signed difference between the given father's height and the mean height of the population.

23. In Note 9.2.2c, how might you test between the two hypotheses (a) experimentally and (b) with observational data?

23ANS. One possible experimental test would be to present tethered insects to each spiderling to guarantee each an initial catch. Measure their latency and correlate this with subsequent successes.

Relevant observational data would be spontaneous activity, or speed of orientation to some alerting event. The constitutional hypothesis implies these measures will correlate with subsequent success.

24. Verify the derivations of Equations 6b, 7b, 13, and 14.

ANSWERS FOR CHAPTER 10

Please do all calculations using only hand calculator. This will help you understand the basic *proportionality rule* for getting expected values.

1. The discussion of the polio data implies that a statsig X^2 is not adequate evidence that the vaccine was effective. Why not? What general principle does this illustrate?

1ANS. The statsig chi-square is needed as evidence for something more than a chance effect, but does not pin down the cause. Control of confounds by using random assignment and double blind procedure were crucial aspects of inference from these data.

The general principle is that scientific inference rests largely on extrastatistical considerations. These considerations require personal research judgment beyond the ken of statistics, which looks only at the numbers and knows nothing about experimental procedure.

2. In the polio experiment:

   a. Why should a preliminary power calculation be made?

   b. What criticism might be made of the N in this experiment?

   c. By analogy to Section 4.3 on power for Anova, guess what item of information is needed for a power calculation for chi-square.

2ANS. a. It would seem unethical to have done this polio experiment without a preliminary power calculation to determine the needed N. Too large an N wastes time and money; too small an N is even more wasteful.

b. A possible criticism is that the N may have been substantially larger than necessary. The vaccine was far from perfect, yet the chi-square was extremely high. In hindsight, at least, a substantially smaller experiment may have been adequate. Some statistical techniques allow decisions at specified interim stages in a planned experiment, so it might have been possible to reach the same conclusion and get a vaccine program underway at an earlier time.

c. The informer needed for calculating power is the expected effect size, or in this case, the minimum effect size worth finding out about. This is analogous to σ_A of Equation 4.3, but calculated using Equation 1 for chi-square. It might seem that a measure of error variance is also needed. This, however, derives from the binomial variance of a proportion, $\pi(1 - \pi)/N$.

3. In one of the final tests of Clever Hans, the mathematical horse of Chapter 8, Hans was instructed to perform a simple arithmetic task, such as adding two numbers. On each trial, Mr. von Osten would whisper one of the numbers in Hans' ear, following which Professor Pfungst would whisper the other number. On some of these trials, one or both men knew the answer; on these trials, Hans got 29 right and 2 wrong. On other trials, neither man knew the answer; on these trials, Hans got 3 right and 28 wrong. (Pfungst, 1965, p. 37.)

   a. What is the null hypothesis in this 2×2 contingency table?

   b. Is chi-square really applicable with these small frequencies of 2 and 3?

   c. Show that $X^2 = 43.66$.

   d. Show that the 95% confidence interval is .84 ± .14.

   e. Is this chi-square test really needed?

3ANS. a. The null hypothesis is that the proportion correct is the same when neither man knew the answer as when one or both knew it.

b. The E values are 16.000 for the two "right" cells, 15.000 for the two "wrong cells." These are more than large enough to allow chi-square. The two small frequencies are observed frequencies; the chi-square assumption refers to small expected frequencies.

e. Neither Mr. von Osten nor Professor Pfungst needed a statistical test. Whether you need one depends on how far your number appreciation has developed.

4. In the field study of smoking prevention in the Appendix to Chapter 3, 19%, 24%, and 27% of the cases in the three treatment conditions had not smoked at the three-month mark. Assume each group had exactly 1000 subjects. Show that $X^2 = 18.26$.

5. A clinical trial (= experiment) extending over several years showed that breast cancer was developed by 216 of 6707 women on placebo control, and by 115 of 6681 women on the drug tamoxifen. Should the Food and Drug Administration approve tamoxifen for prescription by doctors?

5ANS. Whether tamoxifen should be approved by the FDA involves two issues. The first is whether the given data provide adequate evidence that tamoxifen has beneficial effects. The second concerns possible adverse side-effects.

The first question, whether tamoxifen has beneficial effects, has three parts. Most important, of course, is whether the experimental procedure was sound, without serious threats from confounding. Assuming a positive answer to this method question, the next two issues are: Is the effect statsig? Is the effect large enough to be worthwhile?

In the 2×2 contingency table, the expected values for cancer and no cancer with tamoxifen treatment are 165.2 and 6515.8. The corresponding expected values for placebo are 165.8 and 6541.2. These yield chi-square = 31.2 on 1 df. Tamoxifen thus yields a comfortably statsig effect.

Granted a real effect, is it large enough to be worthwhile in practice? With an N over 13,000, a tiny effect of little practical importance could be statsig. By visual inspection, the efficacy of tamoxifen is roughly 115 to 216; the odds ratio is .53. So tamoxifen is definitely helpful, although far from a cure-all. At the very least, further work to improve tamoxifen looks worthwhile.

The FDA—and patients—face an additional question in considering whether tamoxifen should be approved: What are the side effects? Some evidence suggests that tamoxifen may increase deaths from uterine cancer and from lung clots. These side effects, however, appear markedly less than the benefit for women at high risk for breast cancer.

(These data are from *Scientific American*, June 1998, pp. 26*ff*. Use of tamoxifen as hormone therapy following breast cancer surgery is accepted clinical practice, but continuing this treatment too long may be harmful; see S. M. Swain, *Journal of the National Cancer Institute*, 1996, *88*, 1510-1512.)

6. In a field experiment in a nursing home, one group of residents received a treatment that emphasized their responsibility to make their own choices and control their own lives. A comparison group received a treatment that emphasized the responsibility of the staff to care for them and make them happy (see further pages 465*ff*). The hypothesis was that the first treatment would increase the residents' sense of personal control, with beneficial health effects.

"The most striking data were obtained in the death rate differences between the two treatment groups" in the 18-month follow-up—only 7 of 47 subjects in the first treatment had died compared to 13 of 44 in the second treatment, which the authors claimed was statsig, "(*p* < .01)" (Rodin & Langer, 1977, p. 899).

    a. Check their claim.

    b. What is the moral of this exercise?

6ANS. a. $X^2 = 2.85$ on 1 df, which is considerably less than the criterial value of 3.84 for $\alpha = .05$. Rodin and Langer did not use chi-square, but instead an analogous test between the two proportions, 7/47 and 13/44, using an arcsin transformation, concluding that: "*z* = 3.14, *p* < .01," (p. 900). This result disagrees with the chi-square.

I applied the arcsin transformation to these data (Dixon and Massey, 1969, pp. 324*f*) and got *z* = 1.73, rather less than the .05 criterial value of 1.96, but in agreement with the chi-square. Rodin and Langer evidently made a mistake.

b. Claims in the literature need scrutiny. Mistakes happen. A no less important moral is the need to develop number sense—visual inspection should warn that the two cited proportions could hardly be statsig at "*p* < .01."

7. Adult-onset diabetes produces high levels of blood glucose because of deficiencies in glucose metabolism. It was widely held that lowering blood glucose to normal levels would reduce cardiovascular complications. The drug tolbutamide lowers blood glucose to normal levels and was routinely prescribed by many physicians. An objective test of efficacy of tolbutamide would be a large undertaking in a long-term study using double-blind, randomized design. Despite the high cost, an experimental study was considered desirable and it was performed with cooperation of a dozen medical clinics.

In addition to the placebo and tolbutamide groups, two groups received insulin, a standard treatment for diabetes. These two groups were expected to be fairly similar. Each group had a little over 200 patients as subjects. Deaths from cardiovascular disease were expected to run about 5% during the study.

    a. Are the expected numbers of deaths in each cell of the contingency table large enough to justify chi-square?

    b. Suppose the X^2 on 3 df for the overall data turns out to be statsig. What does this tell you? What does this not tell you? (See Section 4.2.)

    c. You wish to compare the tolbutamide group with the two insulin groups, ignoring the placebo group. The two insulin groups are expected to show similar results. Give two possible formats for the contingency table. Which one would you plan to use?

    d. Suppose the University Group Diabetes Program, which performed this study, had asked you to plan the statistical analysis for the data on cardiovascular deaths. What tests would you plan as essential? What supplementary tests, if any, would you consider desirable?

    e. What α level would you advise?

7ANS. a. The overall contingency table should have two rows and four columns. Rows correspond to deaths and nondeaths; columns correspond to the four treatment conditions. With some 200 cases in each condition, the 5% mortality rate yields about 10 expected deaths in each condition, comfortably more than the minimum cell frequencies cited in the text.

b. A statsig overall X^2 on 3 df would tell us that cardiovascular mortality depends on the medication. What it would not do is localize the effect. Since specific hypotheses are clear in the plan of the investigation, the overall X^2 has little pertinence and should not be performed. Instead, the specific hypotheses should be tested.

c. One format represents the three groups (tolbutamide and the two insulin groups) as a 2×3 contingency table, yielding a chi-square on 2 df. The other format pools the two insulin groups to obtain a 2×2 contingency table on 1 df. Since the two insulin groups are expected to be fairly similar, the latter test is preferable; it has greater power and is more informative.

d. Two tests seem essential: (1) Between the tolbutamide group and the pooled insulin groups and (2) between the tolbutamide group and the placebo group. Each has 1 df and addresses a specific hypothesis. A supplementary analysis could compare the two pooled insulin groups with the placebo group.

These 1 df chi-square tests are analogous to the two-mean comparisons of Chapter 4. The two essential tests, it may be noted, are not independent. Both test relevant null hypotheses, however, so the nonindependence is not a serious objection.

Making three tests causes the α level to escalate for the family of three tests taken together. Since all three tests represent a priori planned comparisons, however, there seems no need to adjust the α level (Section 4.2.2).

e. For a decision whether to do a replication for further information, an α level of .05 seems reasonable. For a decision about prescribing tolbutamide in medical practice, a more stringent α level would seem necessary, at least if no other information was available. In short, likely costs and benefits are essential elements of any decision, especially for outcome-oriented studies.

8. In the previous exercise, there were 10 cardiovascular deaths of 215 patients in the placebo group, 26 of 230 in the tolbutamide group, 13 of 223 in one insulin group, and 12 of 216 in the other.

a. Consider two comparisons: tolbutamide with the placebo; and tolbutamide with the two pooled insulin conditions. Guess at the X^2 for each test, relative to the criterial X^2 of 3.84 on 1 df.

b. Calculate chi-square for (a). (For the first, $|O - E| = 7.39$ in each cell.)

c. This tolbutamide study was "received by some critics with a hostility which has no discernible scientific basis" said Cornfield (1971, p. 1676) in his reply to critics (published in *Journal of the American Medical Association*, 1971, *217*, 1676–1687; well worth reading for several reasons). What is your conjecture about the origin of this "hostility?"

d. What is the moral of this exercise?

8ANS. b. The value of X^2 is 6.61 for the comparison between tolbutamide and placebo, and 6.74 between tolbutamide and insulin. Contrary to substantial medical opinion, tolbutamide seems actually harmful.

c. The clue to this question appears in the first paragraph of the preceding exercise, which says that tolbutamide was routinely prescribed by many physicians. The thought that they had been harming their patients, not helping them, would elicit denial and other defense mechanisms.

d. The main moral of this exercise is the importance of applying scientific method to every kind of social issue, not relying solely on expert opinion.

9. You wish to compare three types of dyslexic children on ability to solve a certain kind of verbal problem. Each child receives three problems, which are scored success or failure. You had planned to apply chi-square to the frequencies pooled over all three problems, but now you realize that would involve double counting. Accordingly, you limit the chi-square to just the first problem, but this falls somewhat short of being statsig, p just under .10.

a. You notice, however, that the same trend appears in both other problems so you do a chi-square on each of them. They also yield p-values a bit under .10. "Aha!" you think, "the probability that all three have $p \leq .10$ is $.10 \times .10 \times .10 = .001$, by the multiplication rule of Chapter 0. So this result is evidently real." What is the flaw in this argument?

b. It's awful to have three responses when the chi-square analysis allows only one and wastes the information in the other two. How can you restructure the analysis to use all the data?

c. What alternative response measure might avoid this difficulty?

d. How much information do you get from a problem so easy/hard that all/none of the children solve it?
What follows?

9ANS. a. The flaw in this argument is that the scores on the three problems are not independent but are correlated across subjects. In the extreme case of a perfect correlation, the later trials carry zero new information. But even partial correlation violates independence. Independence is essential because chi-square treats every observation as equally informative (as of course do nearly all statistical analyses).

b. Count the number of successes for each subject to get a numerical score; apply Anova. Although this does not yield a normal distribution, there should be no problem unless there is a strong floor or ceiling effect.

c. One possible alternative response measure is time to do the problem. This may require a cutoff limit on the total time allowed on each problem to handle nonsolvers, or use of a reciprocal transformation. Being magnitude scores, however, response times may carry more information than the 0–1, success–failure scores, even if a rank-order test is needed.

d. A 50% success rate may give optimal discrimination.

10. Show that pooling the three sets of figures on graduate admissions cited in *Gender Bias?* in the last subsection of this chapter yields the cited overall admission rates for males and females.

11. A medical survey of thyroid and heart disease in Whickham, England in 1972–1974 used a one-in-six sample of the electoral rolls. A followup 20 years later obtained the following survival data for all women in the original survey who had at that time been classified as smokers or having never smoked. (After Appleton, French, & Vanderpump, 1996.)

	Smoker	No-smoker
Dead	139	230
Alive	443	502

a. Summarize the sense of this table with two numbers. At face value, what do these two numbers imply?

b. Would it be valid to apply a chi-square test to these data?

c. The observed outcome may seem contrary to other knowledge. What missing variable might account for the observed pattern?

11ANS. a. The sense of this table in two numbers is that 31% of the nonsmokers died, but only 24% of the smokers. Smoking must be good for your health!

b. The chi-square test is valid. It seems likely to be comfortably statsig. As always, the statistical test only looks at the numbers. The obvious implication—that smoking is good for health—may be incorrect, but this is outside the province of chi-square.

c. Age is a missing variable. In the original sample, older women were less likely to have smoked and more likely to have died in the interim. This was borne out in a breakdown by age.

NOTE: Missing variables do not often reveal themselves in this blatant manner. When missing variables do have important effects, these often look reasonable, as in the gender bias example.

12. Suppose the 199 polio cases in Table 10.1 had been distributed 3 and 196. Calculate both ϕ and the odds ratio and compare them with the corresponding values for the 57–142 split given in the text. Comment.

12ANS. With the 196–3 split, the odds ratio is of course much higher, specifically, 65.2. With the 196–3 split, $X^2 = 186$, and so $\phi = .02$.

The odds ratio increased by a factor of 26 (= 65.2 ÷ 2.49); the ϕ index increased by a factor of 2.2 (= .02 ÷ .009). The vaccine looks a lot better with the odds ratio.

Comment: The ϕ index is looking at the overall picture. This ϕ index is low because polio is a rare disease, a fact that is ignored with the odds ratio. For medical research, the odds ratio seems more pertinent. The odds ratio may be misleading for social decision, however, which should take into account the costs and benefits.

13. **Median Test**. Chi-square may be used as a *median test* for data that can be rank ordered. To illustrate the median test, consider the rank order of admission of the 40 patients in the study of controlled drinking listed in Exercise b2 of Chapter 12. Dichotomize the 40 subjects at the median admission time to get a 2×2 contingency table. Apply chi-square to test whether the greater proportion of Control subjects admitted in the second half is real.

13ANS. The first 20 patients were 14 E and 6 C; the second 20 were 6 E and 14 C. The null hypothesis implies Es of 10 in each cell. Hence $X^2 = 4 \times 4^2/10 = 6.40$, which is substantially greater than the criterial value of 3.84.

14. How does *The Case of the Molded Plastic* in Section 8.1.5 illustrate the threat of missing variables?

15. In Section 10.2.1, explain why N is irrelevant to the strength of association. What is N relevant to?

15ANS. a. The strength of association is a property of the population. A larger sample will give a more reliable estimate of the strength of association, but will not affect its size. The same point applies to the correlation coefficient of Chapter 9. N is relevant to reliability and hence also to power.

16.* Cases occasionally arise that look like a two-way contingency table, but require a different analysis. Here is one such case (Karpp & Anderson, 1997).

Two methodologies have been used to study whether peoples's intuitive judgments of physical tasks have the same form as the physical law. One is the Piaget–Siegler choice methodology described in Section 8.1.3 and Note 8.1.3b; the other is the functional measurement methodology of Chapter 21. Do the two methodologies agree in their diagnoses of cognitive knowledge systems?

Subjects judged how fast a cart of specified *mass* released from a specified *height* on an inclined plane would be moving when it reached the bottom. Subjects responded to a set of mass–height combinations constructed to provide the diagnostic information needed for each methodology.

Of the 40 subjects, 3 were diagnosed as centerers by functional measurement, and these 3 were diagnosed the same by the choice methodology. The remaining 37 subjects were all diagnosed as integrators by functional measurement but only 8 were diagnosed as integrators by the choice methodology, the other 29 being diagnosed as centerers.

a. What is the null hypothesis for comparing the two methodologies?

b. Make a 2×2 table with rows corresponding to the two diagnostic methodologies, columns to the two diagnoses (integration, centering). Why would it be inappropriate to apply chi-square considered as a test of homogeneity or of independence (defined in Note 10.1.4a)?

c. What statistical test would you apply?

16ANS. The null hypothesis is that the two methodologies agree, giving equal numbers of each diagnosis. The 2×2 table of (b) cannot test homogeneity because there is only one group. Test for independence is not to the point because the question at issue is whether the two metholodies disagree, not whether they partly agree,

Disagreement between the two methodologies can only show itself in the two cells in which they give different diagnoses. Chi-square may be applied to test whether these two observed frequencies, 0 and 29, are likely to be observed if the true proportions are equal; see *Goodness of Fit* in Section 10.2.1. (Of course, a formal test is hardly necessary to show that 0 and 29 are statsig different.)

This study showed that most subjects integrate in tasks of intuitive physics, which disagrees with results repeatedly obtained by workers who used the choice methodology. Functional measurement methodology revealed algebraic integration rules in single subject analyses, even with young children.

17. ''A given value of $(O - E)^2$ is weaker evidence when E is large than when E is small'' (Section 10.1.2). Intuitively, why is this true? Justify it with a formula from Chapter 2.

ANSWERS FOR CHAPTER 11

NOTE: For the following exercises, assume independent scores even in serial observation design unless otherwise indicated. Independence always needs careful consideration; it is assumed here in order to focus on other problems.

1. You run four single subjects. Three show a substantial effect, comfortably statsig. The fourth shows nothing. You submit these results for publication, including four single subject Anovas. Among other editorial reactions, one reviewer requests a repeated measures Anova for all four subjects. The Editor indicates a revision should be acceptable, without remarking specifically on the repeated measures analysis.

   a. While you're fretting about how to handle the revision, your assistant rushes in, crying, "I've done the repeated measures analysis. Shall I show you the results right now?" How should you reply?

   b. What do you conclude about the reviewer's request?

   c. Suppose you conclude the repeated measures analysis requested by the reviewer is not appropriate. With your revision, you include a letter to the Editor that explains your reasoning. Nevertheless, the Editor makes this analysis prerequisite to publication. Now what do you do?

1ANS. a. Reply "No." How you respond to the editorial request should be decided on principle. If you look at your assistant's repeated measures analysis, the outcome might tempt your principle.

b. In my opinion, the repeated measures analysis is less than worthless. Statistically, it might be argued that it allows a generalization from your sample of four subjects to a population. But you already have such a generalization: Some individuals show an effect; so there is an effect in the population.

   Furthermore, such statistical generalization would be misleading because it refers only to the mean of the population. Your results suggest that this population is heterogeneous so the population mean conceals important information about individual differences. Your single subject analysis is thus preferable in every way.

c. If the Editor insists on including the repeated measures analysis, I would include it with a comment that it was requested in the editorial process but that, for the reasons just noted, you consider the single subject analysis preferable. Don't try to show up or beat down the Editor, who has enough troubles in a demanding, thankless job and should not be expected to be more perfect than anyone else.

2. Construct a *very simple* conceptual example in which an A-B design suffers a total confounding that is revealed with an A-B-A design. Include a numerical, error-free example in graphic form.

2ANS. To illustrate total confounding in an A-B design that is revealed in an A-B-A design, let the response increase by 1 each successive session in each period. Then B is better than A in the A-B design, but this conclusion is shattered in the A-B-A design.

In this simple example, of course, the steady increase over the initial A period is a clear warning about temporal trend that is revealed in the final A period.

   As an alternative, let the response jump from one constant level in the initial A period to a higher constant level fairly early in the B period and remain at the same level in the final A period. Both examples are simplistic, but they illustrate real dangers that appear with real data that are variable and not simplistic.

3. The preface to this chapter, considering whether the A-B difference is reliable in an A-B design, asserts "The answer obviously requires comparison of the mean difference between treatments to the variability within treatments."

   a. Why exactly is *this* comparison *required*? (Of course, "mean" could be replaced by "median," for example, but this statisticality is ignored here.)

   b. Do you think the word "obviously" is justified?

3ANS. The prevailing response variability within each condition will produce some differences between the sample means. Any claim that the difference is reliable should show that it is larger than could reasonably be expected from the response variability.

   This is just the idea of a test of significance, regardless of whether it is made by visual inspection or by Anova.

b. "Obviously" represents a first principle of empirical science—that a sample mean should be viewed as an interval of likely error. Likely error is measured by the variability within conditions. This principle applies equally to visual inspection as to formal statistical tests. In practical affairs, of course, some persons do take data at face value, not having learned the first principle—that the sample mean is an interval of uncertainty.

4. To test relative efficacy of two therapy regimes (drug plus exercise) for a patient suffering chronic pain, a hypothetical patient–physician team used an A-B-A-B design. Each period was long enough to wash out effects of the previous period. Mild serial correlation seemed not unlikely so the response measure was taken as the mean over the last three observations in each period. At the end of the four successive periods, the response measures were 14, 19, 12, 17 (larger numbers mean greater pain).

   a. By visual inspection, guess how close $F(1, 2)$ will be to the criterial 18.5.

   b. Do the Anova and interpret the result.

   c. Suppose this patient is a member of your family, not conversant with statistical analysis. In light of (b), what do you advise?

   d. Assuming this outcome is real, are there any confounding factors that seem a *serious* concern for interpretation?

   e. What, if anything, does this Anova add to visual inspection?

4ANS. b. The means are 13 and 18 for the A and B treatments. The SSs for treatments and error are 25 and 4, on 1 and 2 df. Hence $F(1, 2) = 12.50$, less than the criterial $F^*(1, 2)$ of 18.5. Although not statsig, I would consider that this difference is suggestive. (With so little data, of course, F is not robust against nonnormality.)

c. Tell your family member by all means to use medicine A (side-effects and costs being equal). Since it did better, it should be used regardless of any significance test.

d. A placebo effect might be obtained if the patient got some idea, however it might be obtained, that one medicine was better than the other. It might be difficult to ensure that both medicines had the same odor, color, consistency, and so on. A placebo can be good medicine, of course, if the patient does feel better. With this chronic condition, however, a placebo effect may be rather temporary and not give enduring relief.

Also, extraneous events in the patient's life might have caused a better score in the two A periods, independent of the medicine. This confound could be serious. An additional A-B period would of course have allowed stronger evidence.

e. If you thought the difference was clearly reliable, Anova cautions you that it fell somewhat short. If you thought the difference was clearly unreliable, Anova informs you it is substantially better. If you thought the difference was promising but not statsig, Anova gives you a hearty pat on the back.

5. Of the A-B design, Section 11.3.1 states "Given a long, flat trend under A, it seems unlikely that the behavior would change just when B is introduced unless B had a real effect."

   a. What is the rationale for this statement?

   b. What is the null hypothesis in this test? Be precise.

   c. Can this null hypothesis be false if A and B have identical effects?

5ANS. a. It *is* unlikely that the behavior would change spontaneously just when B is introduced. This is a visual test of significance, based on the long, flat trend, which implies that spontaneous changes are infrequent.

The real question concerns the cause of the change, whether the treatment or some other cause. Two expectations are relevant: The treatment is expected to change the behavior; and the behavior is not expected to change from some normal aspect of the situation or of the subject. In any actual case, of course, we would have extrastatistical knowledge about stability of the behavior and likelihood of external causes.

b. It is incorrect and dangerous to think that the null hypothesis is that treatments A and B have equal effects; that ignores the potential confounding. The null hypothesis refers to the empirical data situation: That the true mean score is the same after as before the B treatment is begun. A statsig difference is not enough; the extrastatistical inference of (a) is needed to conclude a real difference between treatments.

c. Yes; there is the possibility of a confounding factor already discussed in (a).

6. In an A-B-A design with 6 (independent) observations in each period:

a. How do you get a confidence interval for the difference between the mean of the two A periods and the mean of the B period? How many df?

b. What can be concluded from this confidence interval? What cannot?

6ANS. a. Do an Anova on the three periods to get an error term on 15 df. Get the means for A and B on 6 and 12 scores, respectively. Construct a confidence interval for the difference between the A and B means using the formula for unequal n in Note 4.1.1b.

Do not combine the scores into an A and B group and do Anova with 16 error df. The error gains 1 df but is inflated by any real difference between the two A periods (which you may also wish to test).

b. If 0 lies outside this confidence interval, you can conclude that the two true means differ reliably. This A-B-A design is rather weak against temporal confounding, so there is a danger in concluding that this real difference is caused by the treatments themselves.

7. In the numerical example of the 2×3 design in the Appendix to Chapter 5, suppose these scores had been obtained from a single subject, using treatment randomization to obtain independence, rather than 12 independent subjects, randomly assigned.

   a. What changes would be required in the Anova?

   b. What can be said about the generality of the results assuming subjects are (i) a random sample and (ii) a handy sample?

7ANS. a. The Anova is identical for the group and single subject design because the numbers are identical. Both designs thus yield equivalent Anovas.

b. In each case, statsig results may be generalized to the given sample of subjects.

If both samples were random from some larger population of subjects, results from the 12-subject design could be statistically generalized to the larger population. But the results for the single subject could not; there is no measure of individual differences.

In practice, both samples would usually be handy samples, not random. Extrastatistical generalization is still in order but weaker with the single subject.

8. You run a single subject in a simple A-B design. Assume no confounding is present so the A-B difference can be interpreted at face value. How do you decide how many trials to run in each period? How does this apply to the experiment of Figure 11.5?

8ANS. Since you wish to obtain reasonable evidence for a real difference between the two treatments, you should make a power calculation to help decide how many trials to run. This is just a two-group power calculation, which may be done exactly as in Chapter 4. Had such a power calculation been run in the experiment of Figure 11.5, it would surely have warned that power was inadequate.

Of course, you may have further purposes than merely showing a statsig difference. Hence additional trials under one or both treatments may be desired in order to, for example, get narrower confidence intervals.

9. In subsequent studies of the blame schema of Figure 11.3, each of a substantial number of subjects received two randomized replications of the design.

  a. Write out the Anova table (Source and df) for a single subject in this task.

  b. Of what interest are the main effects?

  c. Of what interest is the interaction residual?

  d. What argument can be made for using $\alpha = .10$?

  e. Is visual inspection useful for schema diagnosis in this case?

9ANS. a. Sources are intent, damage, intent $\times$ damage, and error, on 2, 3, 6, and 12 df.

b. Main effects have interest for diagnosing individual children who use only one of the two given informers.

c. The interaction residual has interest for diagnosing the integration rule for those children who use both informers. This point was illustrated in Figure 11.3.

d. Using $\alpha = .10$ here provides more power for detecting children who place major reliance on one informer, minor reliance on the other. With only 12 df for error, which is perhaps as much as could be obtained with young children, the .05 level would tend to misclassify some subjects who actually do use both informers. The .10 level seems reasonable because a false alarm with an occasional child would not be serious.
  In this example, the significance test is used in a qualitatively different way from its standard use in the classic Experimental-versus-Control group design, in which a false alarm is usually costly.

e. Diagnosis of an integration schema depends on the pattern in the factorial graph. As with the accident-configural rule of Figure 11.3.
  A statsig interaction residual only infirms an additive schema, without saying anything about the operative schema. A nonstatsig interaction residual may fail to detect some systematic deviation from parallelism. In fact, nonparallelism is sometimes localized in the linear trend component of the residual, and is washed out in the overall F.

10. P plans a serial observation design in which he expects a small gradual increase in response level over the course of his experiment as well as a substantial difference between effects of A and B. P presents this plan to your research group, saying he intends to use an A-B-A design with equally many trials in each period. He plans to compare the mean of the B period with the pooled mean of the two A periods.

  a. Suppose there is a linear trend over trials. Will this confound P's results? (A numerical example may be helpful.)

  b. Do you think P's design and plan are reasonable?

10ANS. a. Suppose there is no A–B effect, only a linear trend

over trials. Then the mean response in the B period will be higher than the mean response in the first A period. But the mean response in the second A period will be twice as much higher. Averaged over both A periods, therefore, the trend will cancel out. Hence P's comparison will be zero, just as it should be. Any real A-B effect is thus unconfounded with the linear trend.

b. If the trend is nonlinear, P's comparison will not be zero but nonlinear trend is expected to be quite small. If there is a substantial treatment effect, then P's comparison will be substantial. Accordingly, I see no serious objection to P's design. A small effect would be unconvincing, of course, because it could reflect nonlinear trend. An additional B period would provide some protection against nonlinear trend.

Exercises 11a–11b concern the experiment of Figure 11.4; Exercises 12a–12d concern the experiment of Figure 11.5. Although visual inspection seems sufficient for these data, they are used here to illustrate Anova for serial observation design. Serial independence is assumed, as seems not unreasonable since sessions were at least one day apart. Read the data approximately from each figure, and make a neat table. Omit the last point in both sets of filled circle data in Figure 11.4.

11a. a. Do Anova on periods 2 and 3 of Figure 11.4 (with six and three data points, respectively). Interpret this result.

  b. Show that the 95% confidence interval for the mean difference between these two sets of scores has width of approximately ± 15.6 (Note 4.1.1b for unequal n).

  c. Can you guess from visual inspection why the last point of each set of filled circle data was omitted? (Note 11.3.1a.)

11aANS. The actual values, kindly provided by Ted Carr, are $\{8, 15, 5, 21, 10, 25\}$ and $\{87, 65, 66\}$. Reading values from the graph will be more or less inaccurate, but adequate because the differences between conditions are so large.

a. The Anova yields $MS_A = 6884$ on 1 df, $MS_{error} = 87.5$ on 7 df. Thus, $F = 78.7$, far greater than the criterial value of 5.59.

b. The half-width of the 95% confidence interval equals $\sqrt{F^*}$ $\times \sqrt{87.5(1/6 + 1/3)} = 15.6$. The 95% confidence interval is thus 58.67 ± 15.6. This should be rounded to 59 ± 16 because the first decimal digit is clearly unreliable noise.

c. The last point in each set of filled circle data period is zero. It would seem that the subject was being run to a criterion of zero. If so, this last point is biased (Note 11.3.1a).

11b. In Figure 11.4, the use of only three sessions in the second set of open circle data suggests the investigators noticed that the behavior had returned to its original level and decided on this basis to resume the experimental treatment.

  a. What is the advantage of this approach?

  b. What artifact endangers this approach (see Note 11.3.1a)?

  c. How could this danger have been avoided?

11bANS. a. Terminating a phase of treatment when the subject reaches a criterion is a form of interactive design, which can take advantage of the immediate behavior to avoid unnecessary trials.

b. The artifact arises because the last data point is biased by selection, as explained in Note 11.3.1a. In this graph, it seems clear that the bias is relatively small.

c. The possible bias can be avoided by running one or more trials past the trial on which the subject reaches the criterion. The bias is not present in these added trials.

12a. In Figure 11.5, how does visual inspection show not statistical test is needed between placebo and 6-week follow-up (3 and 10 observations)?

12aANS. Visual inspection shows a little lower response under the drug condition.

12b. In Figure 11.5, do Anova for the four dosage levels (placebo, 5, 10, and 15 mg) using the data from the four sets of three data points. Find 95% confidence interval between largest and smallest means. Interpret the outcome.

12bANS. The four sets of three data points are {142, 172, 150}, {168, 156, 222}, {162, 140, 135}, and {152, 184, 160}, for placebo, 5, 10, and 15 mg drug, respectively, read approximately from the original figure. The (approximate) mean squares are 730 for conditions and 490 for error, yielding $F(3, 8) = 1.49$, far less than the criterial F^* of 4.07.

12c. a. Do a power calculation for the difference between the 5 mg and placebo conditions, using the mean square error of 490 from the previous exercise. What does this mean to you?

b. How large a difference between means would be needed to get power of .80 for the comparison of (a)?

12cANS. a. The power calculation is based on Equations 3–5 of Chapter 4. With the two means taken as 182 and 155, the α_j in Equation 4.3 are 13.5 and − 13.5. This yields $\sigma_A = 13.5$. Also, $\sigma_\varepsilon = \sqrt{490} = 22.1$. Thus, $f = \sigma_A / \sigma_\varepsilon = .61$. Finally, $\phi = .61 \times \sqrt{3} = 1.06$ on 1/8 df. The power curve does not go this low, but it shows that power is less than .30. This warning can lead to search for better experimental design (e.g., Exercise 12d).

b. To get power of .80, we read off the power curve for 1/8 df that ϕ must equal 2.26. To increase ϕ from 1.06 to 2.26 would require increasing the mean difference by the ratio 2.26/1.06, that is, from the observed 27 words to 58 words.

12d. Methylphenidate has a half-life in the body of 4-6 hr and is thought to be completely eliminated within 24 hr. On this basis, design an experiment for Bill (Figure 11.5) based on a Latin square (Section 14.3), including pertinent details of procedure. Discuss pros and cons of this design.

12dANS. Given that methylphenidate is completely eliminated from the body in 24 hr, it appears feasible to change the dosage level from one day to the next. The four conditions may thus be applied on one day in each week. Each week could represent one row of a 4 × 4 Latin square, with 4 weeks yielding a complete square. This time requirement is somewhat less than that of the given experiment.

A square balanced for effect of previous treatment (shown in Table 14.7) seems desirable. If there was indeed no carryover from one day to the next, it would be useful to get this on record for future experiments.

This square allows assessment of trend over days within weeks and trend over weeks; these correspond to the position effect and the sequence effect in the Latin squares of Section 14.3. Since there are few df for error, however, it deserves consideration whether to pool these sources with error by a priori decision. The balance provided by the square is present regardless of whether these sources are evaluated or pooled. Although some increase in error may be expected, it may be outweighed by the gain in df (Section 18.4.4).

An alternative would be to use only three conditions, placebo, 7.5 and 15 mg drug. The given data show so little effect that the use of four drug levels seems less than optimal, at least in hindsight. Reducing to three levels would increase the number of observations obtainable for each level in the same time and thereby increase the reliability of the means. To balance for previous treatment requires both possible 3 × 3 squares. This could be done in 18 days, two more than for the 4 × 4 square.

The Latin square is a much more effective experimental design than was actually used.

13. In Figure 11.7, percentage correct responses on successive blocks of 40 trials were as follows for two conditions on day 6:

Old position, old orientation: 83, 88, 85, 85, 83, 78;
Old position, new orientation: 65, 68, 80, 63, 70, 70.

a. Get the error half-bar for a single mean and give its meaning.

b. Get a confidence interval for the mean difference and interpret.

c. What two hints does visual inspection suggest?

13ANS. a. Anova on the two groups of scores yields $MS_{error} = 23.07$ on 10 df. The error half-bar is $\sqrt{MS_{error}/6} = 1.96$. This error bar is the standard deviation of a single mean, yielding an approximately 65% confidence interval.

b. The 95% confidence interval for the difference between the two means is $14.33 \pm \sqrt{2} \times 2.23 \times 1.96 = 14.33 \pm 6.18$. Since 0 lies far outside this interval, we may conclude that changing the orientation of the stimulus lines by 90° causes marked decrease in discrimination.

The increase in discriminability, therefore, failed to transfer even to the same stimulus at the same retinal location but with a different orientation. This result reinforces the main theme of this study that this perceptual learning was retinally localized.

c. The new orientation seems to produce more variable data. Also, there is some sign of a peak in the middle of the session for both orientations. These are slight hints, of course, but they deserve to be checked out with other data.

14a. a. How does Sternberg's additive-factor method of Section 11.4.4 avoid the cited shortcoming of Donders's method (see also Exercise 6.19)?

b. The main effects are obvious. Don't belabor them with a significance test. But say why a small main effect would be undesirable.

c. How does the nonadditivity of Number of Alternatives with Stimulus Quality and with Stimulus–Response (S-R) Mapping support and buttress the interpretation that the latter two variables are additive?

d. Comparison of the RTs with two and eight alternatives might be safest if based on only the same numerals from the latter as from the former. Which two numerals would you use in the two alternative condition?

e.* The two-way nonadditivity of Number of Alternatives with S-R Mapping was 80.9 ms—over three times larger than two-way nonadditivity of Number of Alternatives with Stimulus Quality. Why does this difference not imply that Number of Alternatives is more important for the S-R Mapping process than for the Stimulus Quality process? (See *Concept–Instance Confounding* on page 232.)

14aANS. a. Sternberg's additive-factor method avoids the shortcoming of Donders's method in two ways. In principle, it has better prospects because it only seeks to manipulate the time required for given stages; Donders's method intrudes an alien stage, which may change the nature of the entire sequence of processes. In practice, Sternberg's hypothesis implies zero interactions in Anova, thereby allowing a test of its validity. It does not assume its own correctness, as Donders's method did.

b. A substantial main effect is necessary to show that the target process has been manipulated. Furthermore, any nonadditivity will generally be a fraction of the main effects, so a small main effect would indicate inadequate power to detect nonadditivity.

c. Finding the two cited nonadditivities of Number of Alternatives gives general reassurance that the experimental conditions had enough power to detect nonadditivity. In particular, the two nonadditivities indicate that Stimulus Quality and S–R Mapping both had sufficient range to reveal mutual nonadditivity.

d. For the two alternative condition, I would avoid using numerals 1 and 2 on the ground that the "plus 1" response would be stronger for these from everyday counting. Also, I would avoid 8 as this is the last in the series. Differences between numerals could be assessed with the practice data before fixing final details.

e. This question reemphasizes that manipulation of some variable must be done with specific, concrete levels (see *Concept–Instance Confounding* in Section 8.1.4). The results may be limited to these specific, concrete levels, not applying to the variable per se. The variable per se is confounded with its specific levels; a larger range of Stimulus Quality might have yielded an opposite outcome. This is a sometimes insuperable barrier to measuring importance or comparing importance of different concepts (Section 18.1).

14b. Latin square design seems to be underutilized in perception/cognition. Suppose the Latin square of Table 14.3 is used for a single subject in Sternberg's additive-factor experiment of the previous exercise. Consider just the case of two alternatives. The 2×2, Stimulus Quality $\times$ S-R Mapping design then has four conditions, which are assigned the letters A_1, A_2, A_3, A_4, in Table 14.3. In a single session, the subject receives a block of trials at each of these four conditions, given in the order specified by the first row of the square, next in the order specified by the second row of the square, and so on. The primary response is the mean reaction time over a terminal subblock of each block of trials, the initial trials being omitted as warmup adjustment to the given condition.

The Anova at the right side of Table 14.5 may be applied, with $a = g = 4$ and $n = 1$. With a single subject, of course, some sources will have different meanings than indicated in the discussion of Table 14.5 for different subjects.

a. Position effects refer to progressive changes in response level over successive trials. What two sources in the Anova table at the right of Table 14.5 would represent position effects in the given Latin square?

b. What would be the likely origin of a position effect? How large would you expect position effects to be in Sternberg's experiment?

c. Suppose treatments have no effect but that the response level is 1 on the first trial and increases by 1 on each successive trial. What is the numerical relation between the Position and Sequence effects? Would you expect the same qualitative relation to hold generally?

d. The four treatments form a 2×2 design so $SS_{Treatments}$ on 3 df can be broken down into three parts. What do these three parts measure?

e. The AG term in the right panel of Table 14.5 corresponds here to the interaction of treatments with Sequences (rows). This term is used as the error term for all sources (including Sequences), that is, as an estimate of the response variability of this subject under the same experimental conditions. Why is this error term expected to be a little too large?

f. Suppose the experiment was run in four sessions, with just one row of the square in each session. What, if anything, would change in the analysis?

g. Why can $MS_{SA/G}$ be used to test Sequences?

14bANS. a. One locus of position effects in the Anova of Table 14.5 is obviously the Position effect, which represents changes in reaction time across the four columns of the Latin square. The other is the Sequence effect, which represents differences between the row means; the row means are averaged over the same four treatments for the given subject, but will differ if reaction time changes systematically over trials.

b. Practice–learning–fatigue seems the most likely origin of any position effect. These should be pretty small because of the extensive preliminary training.

c. The Sequence effect should be the most sensitive measure of position effects. In the given numerical example, the differences between Sequence means will be 4 times larger than the differences between Position means. The same direction of difference will be obtained as long as the position effect has the same directional effect at each position.

d. The four treatments form a 2×2, Stimulus Quality $\times$ S-R Mapping design. Hence the 3 df for Treatments in Table 14.5

may be broken down in the standard way into two main effects and interaction residual, each on 1 df.

e. Ideally, each treatment would have the same effect in each row. Then the AG, Treatment × Sequence interaction would have the same expected value as MS_{error}. The same holds if each row adds a constant to the treatment effects. More likely, of course, is that position effects are not constant from one row to another, which nonconstancy will appear as Treatment × Sequence interaction. This may be expected to be quite small, however, in view of the extensive preliminary training.

f. The Anova remains identical but the Sequence effect includes day variation.

g. The same subject is in all sequences, whereas different subjects are in each sequence in the applications considered in the discussion of Latin squares in Chapter 14.

15. In Schlottmann's study of utility theory in Figure 11.9, the adult group did not violate utility theory; for them the dashed curve lay below and nearly parallel with the solid curves. Accordingly, the disordinal violation of classical utility theory in Figure 11.9 might be considered immature behavior by children. How do you think Schlottmann answered this question in her Discussion?

15ANS. Developmental psychologists know that the "child is father to the man" (and mother to the woman)—that understanding adult behavior rests on understanding its course of development. In general, therefore, understanding the well-known biases in adult judgment–decision may depend on tracing out their developmental origins. In particular, the ubiquity of the averaging rule in adult cognition suggests that the addition rule observed by Schlottmann may be limited to relatively simple situations.

16. In a 4 × 4 study of perception of cold temperature, you give each treatment once to a single subject. You had planned to replicate, but this experiment is arduous and time-consuming, and the subject could only with reluctance be persuaded to complete the first replication.

a. How can you get an approximate significance test for the main effect of each variable?

b. What disadvantage does this test have? How serious do you consider it?

16ANS. To test main effects with $n = 1$, the interaction on 9 df can be used. It is conservative to the extent that interaction residuals are real, but it would often serve adequately.

17. In the first paragraph of this chapter, what justification can be found in the Anova model of Equation 2 of Chapter 6 for the assertion that "error variability will be even less than with repeated measures design"?

17ANS. The error term for repeated measurements design in Equation 6.2 has two components: The subject–treatment interaction residuals [the $(S\alpha)_{ij}$] and the within-individual variability component [the ε terms]. The former component disappears with a single subject, leaving a smaller error.

ANSWERS FOR CHAPTER 12

a. Trimming

a1. Consider the numerical example of trimming in Table 12.1.

    a. By visual inspection, show that A_1 and A_2 have equal trimmed variance.

    b. By visual inspection, how will the trimmed mean for A_2 change if the 24 is changed to 24,000,000?

    c. By visual inspection, how will the width of the confidence interval for the trimmed mean for A_2 change if the 24 is changed to 24,000,000?

    d. Use the calculations given in the table to show that the standard deviation of a trimmed mean is 1.14.

a1ANS. a. Inspection of Table 12.1 shows that the trimmed A_2 data equal the trimmed A_1 data plus 4. Since adding a constant does not change the variance, they must have equal variance.

b. The largest score for A_2 will be trimmed to 10, regardless of its numerical value. Its numerical value thus has no effect on the trimmed mean. With this monstrous outlier, of course, F_A on the raw data would virtually vanish.

c. Since the trimmed variance does not depend on the numerical value of the largest A_2 score, neither does the width of its confidence interval.

d. The standard deviation of the trimmed mean is the square root of the trimmed variance divided by the trimmed sample size: $\sqrt{6.50/5} = 1.14$.

a2. In Table 12.1, suppose three scores were trimmed in each tail.

    a. In the trimmed Anova, which calculations would remain the same? Does this hold in general, or is it peculiar to these samples?

    b. What calculations would change?

a3. The following scores represent number of successful responses in 30 trials on verbal concepts by two groups of 7 dyslexic children, obtained by Q in her thesis research.

0	0	11	12	13	14	15	16
4	5	6	7	8	9	10	11

    a. Get F for the untrimmed data.

    b. Trim two scores in each tail and show that $F = 10.71$.

    c. Get the 95% confidence interval for the mean trimmed difference.

    d. What major empirical questions are raised by the two scores of 0?

a3ANS. c. The confidence interval for the mean difference uses $t*$ on 6 df, trimmed MS_{error} of 4.667, and trimmed n of 4 to get 5 ± 3.74.

d. First, the two scores of 0 suggest that the mean superiority of the first group is far from true for some subjects in this group.

Second, these two 0 scores may perhaps reflect some shortcoming of procedure, or perhaps two different kinds of subjects.

In addition, these two 0 scores emphasize that a priori decision about trimming, before seeing the data, is a necessity. Afterwards, it is too easy to rationalize a trimming proportion nudged to fit the data.

a4. Q was uncertain about what trimming proportion to use. So she tried all three of .10, .20, and .30. All three yielded a barely statsig F.

    a. What should Q do?   b. What should be Q's main concern in planning any replication?

a4ANS. Q has a marginal result. A good workman, she will no doubt replicate. A major concern in her future work will be to increase power.

a5. Draw a graph of a two-humped symmetrical distribution to show why the trimmed and untrimmed means have equal expected value.

b. Rank Tests.

b1. Two teaching techniques for parents of autistic children are compared in the graph of Exercise 3 of Chapter 3. Get ranks for the two groups from this graph and do rank Anova.

b1ANS. The graph shows the data in rank order array, so ranking is straightforward. Mean ranks are 16.333 and 8.667. Anova of the ranks yields $F = 9.73$ on 1/22 df. This is rather larger than the F of 8.52 of the raw data. In slight part, this is due to a tie, which is neglected here. Since there seems little reason to question equinormality, the comparison of these Fs indicates that a marginal F_{ranks} should be treated with caution.

b2. The report claiming success of controlled drinking in Section 8.1.6 said that subjects had been assigned at random to the experimental and control groups. These subjects were recruited over a 10-month period as they entered the hospital. Here is the list of subjects in order of admission (e and c denote experimental and control):

    ceeeeceeeeceeeeceeeecceeeecccceeccccccccceccccc.

    a. By visual inspection, do you notice any apparent nonrandomness?

    b. What null hypothesis will be tested by rank Anova? How exactly can the test of this null hypothesis assess departures from randomness?

    c. Show that $F_{ranks} = 13.44$ on 1/38 df.

    d. In light of given information, list one concrete reason why lack of random assignment might bias the results.

    e. How would you have handled the random assignment?

b2ANS. a. Of the last 15 subjects, 13 are Cs.

b. Rank Anova on the order of admission tests the null hypothesis that E and C patients have equal true mean rank of admission. This null hypothesis is violated if the greater observed frequency of Cs in the second half is real.

c. This value of F_{ranks} is a little too large because two pairs of subjects were admitted on the same day and so would have tied ranks, here ignored for simplicity. In a controversial issue like this, of course, exact tables for Mann–Whitney U test would be used to avoid objection, as was in fact done by Pendery, et al.

This case illustrates the importance of careful method and procedure, especially with investigations of high social significance.

d. Lack of random assignment could bias the results because the patient population may well change over the 10 months of the experiment; seasonal changes are a well-known example. If the subjects in the first half were less severe alcoholics, the given order of assignment would bias the results in favor of the experimental group.

e. I would use temporal blocking (Section 14.2). For example, two of each block of four successive patients could be randomly assigned to the E and C treatments. Make up a random subject assignment sheet before the experiment begins; do not trust to flipping a coin as each new patient arrives, as seemingly was done in this experiment.

b3. *Ranks for tied scores.* Tied scores are assigned the mean of their ranks. First rank all scores from 1 to N; for tied scores assign successive ranks. Then replace these ranks for each group of tied scores by the mean of their ranks. If two tied scores receive ranks 2 and 3 in the first step, assign both a rank of 2.5. If three tied scores receive ranks of 7, 8, and 9 in the first step, assign all three a rank of 8. **Check** that the sum of the final ranks equals $N \times (N + 1)/2$.

The following data illustrate this technique for handling ties. (The N is far too small to justify a rank Anova in practice.)

$$A_1:\ 6\ 9\ 9\ 7;$$
$$A_2:\ 4\ 7\ 6\ 4;$$
$$A_3:\ 2\ 4\ 3\ 3.$$

a. Show mean ranks are 10, 6.75, and 2.75 for the three respective groups.

b. Show that $F_{ranks} = 14.17$.

b3ANS. a. The ranks are {7.5, 11.5, 11.5, 9.5} for A_1; {5, 9.5, 7.5, 5} for A_2, and {1, 5, 2.5, 2.5} for A_3. (This too-small set of data is included only to illustrate how to handle ties.)

c. Outliers.

c1. P and Q were now living together, which caused them to think about family dynamics. Seeking experimental tasks to study these phenomena, they decided to compare two praise techniques that one marital partner could use to decrease duration of quarrels. Their pilot work showed one extreme duration, and they wondered how to handle such extreme scores in their main experiment. Q argued that these extreme scores were outliers and that they should use a screening measure to eliminate couples with distressed marriages. P argued that they should use trimming.

a. Discuss justification of each design alternative.

b. What procedure, not necessarily either of these, would you argue for?

c1ANS. a. Q justified the cutoff on the ground that marital quarrels naturally come in two main kinds, those that come and go in working marriages and those that reflect chronic distress by one or both spouses. Although both kinds deserve study, the second is far more demanding. Accordingly, they should adopt the more limited and more feasible goal of concentrating on the everyday quarrels.

P argued that their limited amount of pilot data did not define a cutoff very well. With trimming, it is not necessary to make an arbitrary decision about the cutoff value.

Q replied that in view of the paucity of experimental work on marital quarrels, it was preferable to limit their initial investigations. Furthermore, trimming cuts off the very short quarrels also, which presumably belong to the population.

b. I anticipated that trimming would be the choice for this exercise. After listening to Q, however, I incline to her argument for screening out couples with distressed marriages. In addition, block design on marital satisfaction/dissatisfaction seems desirable (Section 14.2).

A log transformation could normalize extreme scores without trimming the short durations, but I incline toward Q's argument about different qualities of marital quarrels.

This exercise also illustrates the general truth that any course of action will have shortcomings and limitations.

c2. You have 10 difference scores and wish to test the null hypothesis that the true mean is zero. Here are the scores:

$$-2, -1, 1, 1, 2, 2, 3, 3, 4, 5.$$

a. Run regular Anova to test $H_0: \mu = 0$.

b. Show by trial and error that increasing the highest score, 5, to some larger number will (i) increase the size of the effect and (ii) lose the statsig F found in (a).

c. Discuss two implications of your demonstration.

c2ANS. a. $F = 7.01$ on 1/9 df; $F^*(1, 9) = 5.12$.

b. If 5 is changed to 10, the mean increases, but the error variance increases even more to yield $F = 4.95$. This change to one not too extreme score wipes out the statsig result.

c. One implication is that extreme scores can be deadly. A deeper implication is that the size of the effect in terms of the means may not mean much without the error variability. The mean is thus best considered a range of uncertainty, as emphasized in Chapter 2.

d. Transformations.

d1. Consider the following data for two groups:

$$A_1:\ \{1.67,\ 2.00,\ 2.50\};$$
$$A_2:\ \{3.33,\ 5.00,\ 10.00\}.$$

a. Show that $F_A = 4.03$.

b. Apply a reciprocal transformation; find F. Comment.

d1ANS. The reciprocal transformation equalizes the variance for both groups. For the transformed data, $F_A = 13.50$, greater than the criterial $F^* = 7.71$. In contrast, the F of 4.03 from (a) is not statsig, even disregarding the df adjustment for unequal variance.

This example exaggerates the usefulness of transformation, in part because such large heterogeneity would seldom be seen in actual experiments.

d2. Consider the data of Table 7.1.

a. In light of Table 7.2 and Exercise 7.2, why might Q expect more power on main effects than P?

b. In light of his theoretical interpretation, why might P argue that his measurement scale is preferable even if it has less power?

d2ANS. Q's data are parallel so all her SS is concentrated in the main effects, with none in the interaction as in P's data. Greater power would thus seem expected.

Power also depends on error variability, which differs for the two. As a rule of thumb, time scores are skewed and speed scores more normal, which also favors Q. At Ohio State University, Eileen Beier had her clock faces overlaid to read directly in speed instead of time.

b. P could argue that the main effects are built in and trivial, and that the rat's activity takes place in time, which makes time the natural scale. Hence it is the interaction that has main interest. Most experiments are concerned with main effects, he admits, for which Q's speed scale could be preferable.

d3. This and the two following exercises aim to help you develop foresight about what shape distribution to expect in your data. Draw a graph showing what shape of distribution you would expect in the following situations.

a. Random samples of one digit between 0 and 9, inclusive.

b. Random samples of size 5 from the 10 digits, 0 to 9, and take the mean.

c. Random samples of size 3 from the 10 digits, and take the largest.

d3ANS. a. A flat, uniform distribution, with probability of .1 for each digit.

b. By the central limit theorem, these means will exhibit a fairly normal distribution with mean 4.5.

c. This will be a skewed distribution with a long left tail.

d4. Sketch the shape of the distribution you expect for:

a. A final exam in second undergraduate course in research methods.

b. Number of trials to learn a 12-item, English–Spanish paired associates list by undergraduates with no Spanish background.

c. Running times for hungry rats learning a T-maze for food reward.

d. Reaction time in a "vigilance task," in which subjects monitor for very infrequent danger signals over extended periods of time.

e. Age of woman at her first marriage.

d4ANS. a. This course is not usually required, and will mainly consist of motivated and capable students. For most instructors, the grade distribution will have a mean in the upper range, skewed toward lower grades.

b. My guess is for a fairly normal distribution, with a mild tail in the direction of slower learners. Your guess may be better.

c. My expectation is a mode for fast times, with a secondary mode of longer times from trials on which rats took the wrong turn and had to retrace.

d. Occasional long reaction times are expected in vigilance tasks because the subject's attention will wander off. (This can be serious problem in some practical situations, as in highway driving or in military combat.)

e. I haven't checked this distribution, but I expect a mode around age 20, with a long rightward tail. There is also the problem of where to include those who never marry. One solution is to slash the horizontal age scale, and add a single bar at the right end to represent this frequency.

d5. Many distributions do not follow the bell-shaped normal curve. Draw a rough curve—well labeled—for a nonnormal distribution, together with your rationale, for some quantity from each of the following:

a. Everyday life. b. Psychological research.

c. Biology. d. Astronomy.

d6. Nonpositive numbers cause a problem for some transformations. A square root transformation, for example, is not meaningful with negative scores. Similarly, the log of 0 is $-\infty$, not a score anyone wishes to encounter in their data. Suppose you see that a transformation would be helpful but you have a few forbidden nonpositive scores. How can you apply a preliminary linear transformation to eliminate them?

d6ANS. To handle forbidden nonpositive scores with a transformation, first add a constant c to make all scores positive. Then transform $Y + c$. The result depends on the value of c; common practice is to use a near-minimal value of c. With a few zero scores, for example, you could apply the log transform to $Y + 1$.

e. Unequal Variance.

e1. In the third paragraph of Section 12.5, verify that the three standard deviations stand in the ratio $\sqrt{3} : \sqrt{2} : 1$.

e1ANS. The standard deviations for the mean differences are:

for A_3 and A_4: $\quad \sqrt{3/n + 3/n} = \sqrt{6/n}$;

for A_2 and A_4: $\quad \sqrt{1/n + 3/n} = \sqrt{4/n}$;

for A_1 and A_2: $\quad \sqrt{1/n + 1/n} = \sqrt{2/n}$.

These stand in the ratio $\sqrt{3} : \sqrt{2} : 1$.

e2. In Equation 10 for effective df with unequal variance, suppose $n_1 = n_2 = n$.

a. What do you think the df ought to be if $\sigma_1 = \sigma_2$? Verify your answer.

b. Suppose $\sigma_1 \to \infty$. What will happen to df'? Give your intuitive rationale.

ANSWERS FOR CHAPTER 13

1. In the numerical example of *Ancova Idealized* of Section 13.2.1:

    a. Draw a graph to show how "the Ancova adjustment increases $\overline{Y}_E$ by 2 points, decreases $\overline{Y}_C$ by 2 points."

    b. Do the same assuming treatment E increases Y by a constant c for each subject.

    c. Redo (a) using Equation 4.

1ANS. The graphs follow the description of Section 13.2.1.

2. a. Repeat the graphic analysis of part (a) of the previous exercise to show how Ancova would reach wrong conclusions if the estimate of b_1 was lower than its true value, say $b_1 = .6$. Estimate graphically the apparent difference between E and C, and check using Equation 4. How does this relate to evaluation of Head Start programs?

    b. Repeat (a) assuming that treatment E adds 1 point to each subject's score.

2ANS. a. With b_1 less than its true value of 1, the regression line for each group will pass through the middle data point for that group, above the lower data point and below the higher data point. At $\overline{X}_{..}$, therefore, the line for group C will lie above the line for group E, falsely implying that the experimental treatment for group E was harmful.

b. For $b_1 = .6$, Equation 4 yields adjusted means of 3.2 and 4.8 for the E and C groups, respectively. Even if the Head Start program has a positive benefit, the unreliability in the estimate of the true regression coefficient may make it look harmful.

3. The error variances for Anova and Ancova, denoted σ_ε^2 and $\sigma_{\varepsilon'}^2$, respectively, obey the relation (Winer, et al., 1991, p. 741):

$$\sigma_{\varepsilon'}^2 = \sigma_\varepsilon^2 (1 - \rho^2) \frac{df_{error}}{df_{error} - 1},$$

where ρ is the Y–X correlation. For moderate df, the df ratio at the right is close to 1. Ignore this term and plot a graph of the ratio of the two error terms as a function of ρ. How would you describe this trend? What does this mean for experimental design?

3ANS. The graph shows a quadratic trend as ρ varies from 0 to 1, with rapidly increasing decrements as ρ approaches 1. One implication for experimental design is that a correlation as large as .30 reduces MS_{error} by only 9%, which may not be worthwhile unless data are scarce or expensive. With a correlation above .50, however, Ancova can be very beneficial.

4. To calculate power for Ancova, use Equations 3-5 in Chapter 4 for Anova. Replace the error variance, σ_ε^2, in Equation 4.4 by the Ancova error variance from the previous exercise.

    P wishes to study emotional arousal in three kinds of stress situations, but is uncertain how many subjects to run. He guesses the true means are 1, 2, and 4, with an error variance of 6. He plans to measure arousal before and after the experimental stress, the before measure to be used as a covariate. Tentatively, P considers $n = 9$.

    a. Show that power is about .57, .64, .70, and .78 for $\rho = 0$, .3, .45, and .6. Comment.

    b. Guesstimate similarly the 95% confidence interval for difference between two means for each of the four correlations in (a). What problem do they raise for P?

    c. What changes in the experimental design might P consider?

4ANS. b. The 95% confidence intervals for a difference between two means are ± 2.38, 2.28, 2.13, and 1.91, for $\rho = 0$, .30, .45, and .60. Although the difference between means 1 and 4 would be statsig even with zero correlation, the difference between means 2 and 4 would be statsig only for the largest correlation. The difference between means 1 and 2 is not close to statsig in any case. Accordingly, the problem for P is low power to discriminate the middle group.

c. One possible change for P is to omit the middle group. Another is to check power with $n = 16$. The best solution, of course, would be to find some way to reduce the error variability. Perhaps a repeated measures design could be used.

5. In *Blocking Versus Ancova*, justify the statement that the resting state "would probably yield a lower Y–X correlation" if measured in an initial session before the experiment has been started.

5ANS. The subject's resting state is not constant, but will fluctuate or drift from day to day. If resting state is measured for all subjects before beginning the experiment, some days will presumably elapse before each one is run in the regular experiment. Their resting state at that time will then have changed from the measured value, which will thereby have a lower correlation than if resting state had been measured just before starting the experiment for each subject.

6. Here are X and Y scores for two groups, each with four subjects assumed to have been assigned at random.

    group 1: $X = 0, 2, 4, 6;$
                    $Y = 2.8, 4.2, 5.2, 5.8.$

    group 2: $X = 2, 4, 6, 8;$
                    $Y = 3.0, 4.4, 5.4, 6.0.$

a. Which has the greater $\overline{Y}$?

b. Plot both sets of data and fit a separate regression line to each by eye, as illustrated in Figure 9.1.

c. By the randomization assumption, it makes sense to use Ancova. You can do this by eye: Slide the $\overline{Y}$ for each group along its own regression line until its $\overline{X}$ coincides with the overall $\overline{X}$ for both groups taken together. The adjusted Y means for each group can now be read from your graph.

d. Check your adjusted means from (c) against Equation 4.

6ANS. a. Group 2 has the greater $\overline{Y}$, 4.7 versus 4.5 for group 1.

b. The data for group 1 are the same as in Figure 9.1, where the regression fitted by eye had a slope coefficient of 1/2. Group 2 has the same slope.

c. Slide $\bar{Y}$ for each group along its own regression line until it lies above $\bar{X}_{..} = 4$, the mean X for both groups combined. The elevations then become 5.0 for $\bar{Y}_1$, 4.2 for $\bar{Y}_2$. Instead of being inferior by 0.2, group 1 is now superior by 0.8.

Although group 2 was superior on Y, this stemmed from chance random assignment that gave these subjects higher Xs— and thereby higher Ys. Ancova exploits the X information to adjust the correlated part of the random chance in the Ys.

The main statistical benefit of Ancova, it should be reemphasized, is to reduce the error term. The adjustment in means is a lesser benefit, here much exaggerated.

d. With Equation 4, the adjusted means are $4.5 - \frac{1}{2}(3 - 4) = 5.0$ and $4.7 - \frac{1}{2}(5 - 4) = 4.2$, for the two respective groups.

7. The general formula for variance of a difference score is

$$\sigma^2_{Y-X} = \sigma^2_Y + \sigma^2_X - 2\rho\,\sigma_Y\sigma_X,$$

where ρ is the $Y-X$ correlation. Suppose Y and X have equal variance, $\sigma^2_X = \sigma^2_Y = \sigma^2$, as might be expected with before and after measures of the same quantity.

    a. Show that $Y - X$ and Y have equal variance when $\rho = .5$.

    b. Suppose that ρ is greater than/less than $\frac{1}{2}$. Which yields lower variance: the difference score or Y alone?

    c. Why exactly is it preferable to apply Anova to whichever of Y and $Y - X$ has smaller variance?

    d. Why would Ancova with X as a covariate generally do better than Anova of the difference score?

8. In the before–after paradigm, some response is measured on equivalent scales before (X) and after (Y) treatment is given. You use this before–after paradigm to study belief change with two randomized groups, 1 and 2, using the difference score, $Y - X$. But your assistant brings you an Anova on the Y score; it seems the X scores have disappeared. You wonder whether the Anova of Y is valid; perhaps the random assignment chanced to put more subjects with high X scores in one of the groups, a possibility that cannot now be checked.

    a. What null hypothesis is tested by your assistant's Anova of Y? Give answer in words and in symbols.

    b. What null hypothesis would be tested with an Anova of $Y - X$? Give answer in words and in symbols.

    c. Why are the two null hypotheses of (a) and (b) equivalent?

    d. Suppose your treatment has no effect. Do the two Anovas have the same false alarm parameter, or is one better?

    e. Suppose your treatment has an effect. Which Anova do you think has more power?

    f. What advantage would Ancova have over the change score?

8ANS. a. Anova of the after score, Y, tests the null hypothesis that the true Y means are equal: $H_0: \mu_{Y1} = \mu_{Y2}$.

b. Anova of the difference score tests the null hypothesis that true means of $Y - X$ are equal: $H_0: \mu_{Y1} - \mu_{X1} = \mu_{Y2} - \mu_{X2}$.

c. The two null hypotheses of (a) and (b) are equivalent because the true X means are equal by random assignment: $\mu_{X1} = \mu_{X2}$. These two means cancel in the second null hypothesis, which then is identical to the first.

d. Both tests have identical false alarm parameters. One is no better than the other if your treatment has no effect.

e. Which has more power depends on the $Y-X$ correlation (see Exercise 7) and also on the variances of the two groups. As a rough rule, Anova of the difference score has less power. In most applications, Ancova is markedly better.

f. Ancova will nearly always provide a lower MS_{error}. This means narrower confidence intervals and greater power.

9. Why do nonrandom but not random groups suffer from incapacity of one-dimensional covariates to represent multi-dimensional constructs?

9ANS. With random groups, whatever is not represented in a one-dimensional covariate is randomized into the error variance. With nonrandom groups, whatever is not represented in a one-dimensional covariate tends to be confounded with the group comparison.

10. Section 13.2.3 asserts that it is not generally meaningful to ask how nonrandom groups would behave if they were different from what they are. Why does this assertion not apply to random groups?

10ANS. The difference between random groups prior to treatment is due solely to the randomization. The groups could have been different with a different randomization, so it is meaningful to consider the response with a different randomization.

The matter is very different with nonrandom groups. We have far to go on predicting how people will behave given that they are the same as they are (Note 16.1.2b); little can be expected from predicting how they would behave if they were different from what they are.

Dealing with nonrandom groups is a central problem for society. But invocation of some mantra of "statistical control" will not solve them.

11. How does the numerical example in *Bias in Estimating* β relate to the Ancova comparison between nonrandom groups in the Head Start research?

11ANS. This bias means that Ancova gives a false adjustment of the means of the Head Start and the "control" group. This bias can make the poorer treatment look better (Exercise 2), and just this is expected with the Head Start test. As Campbell pointed out, the psychologists in charge of this planning did not seem to know their business.

12. In the cited experiment on premature infants, how might it be helpful to use a preliminary screening test? List potential benefits and disadvantages.

12ANS. The main concern, in my opinion, would be to screen out marginal cases from the main experiment. I have no expertise on this matter, but I guess that cases at the low end are less likely to show much difference from diet and likely to increase the error variance. Low end cases might also be excluded because they require special treatment, but even the low end of includable cases might still be screened out so as to reduce the error variance in the main experiment. An obvious disadvantage is less generality.

ANSWERS FOR CHAPTER 14

NOTE: For the first group of exercises, labeled "a," assume independent scores.

a1. What should be done to avoid possible hurt feelings and distress by a human subject who is eliminated from the experiment for failing to meet some screening criterion?

a1ANS. It seems usually advisable to avoid subjects' learning that they have been screened out; they might feel they have been judged deficient in some way with consequent distress. Running them in the regular experimental task may give useful information about the effectiveness of the screening procedure. Alternatively, they could be given an abbreviated form of the task.

a2. Under *Methods for Language Learning* (Section 14.2.2), explain the sentence, "This was verified, F_{blocks} being even a little larger than $F_{methods}$."

a2ANS. Methods and blocks are both two-mean comparisons on 1 df with the block effect a little larger. Since methods was statsig, blocks must also be statsig. This verifies that the blocks variable was relevant to the behavior.

a3. Q included successive, temporal blocks in her design and found $F_{blocks} = 2.67$ on 3/16 df. What will she make of this?

a3ANS. Q should inspect the pattern of block means for possible clues. The given F falls short of being statsig at $\alpha = .05$, but it is large enough to raise suspicion. Note that a less stringent α level is appropriate for this kind of question.

a4. P used block randomization in six successive, temporal blocks in an experiment in which previous work clearly indicated that block effects were not substantial. P did not include blocks in his factorial Anova of the main treatment variables. By a priori plan, however, he did test the main effect of blocks, using the hand formulas of Section 3.2.5 to get SS_{blocks}, which he tested against MS_{error} from the overall Anova.

    a. What are two objections to P's test of the block effect?

    b. What would P say about these two objections?

    c. Suppose P found a large effect of blocks. Would that require any qualification of his main analysis?

    d. Do you think P erred in not including blocks in his factorial Anova?

a4ANS. a,b. One objection to P's analysis is that ignoring the blocks factor causes block effects to go into MS_{error}. The error term will thus be larger than it should be.

To this objection, P would point out that the block effect is expected to be small by previous studies. Hence this objection does not seem serious.

Another objection is that a linear trend of the block effect would be expected to be more informative than the overall F for blocks.

To the second objection, P should acknowledge it is correct. I trust that P could rightly add that he always looks at the pattern of means and would have noticed a trend regardless of the overall F for blocks.

c. In most situations, blocks is a minor procedural variable, of no great interest in itself and with little relevance to the treat-

ment effects. Finding a substantial main effect of blocks thus seems unlikely to cause any difficulty. Furthermore, possible block–treatment interactions would probably not require serious qualification of the main effect of treatments (Chapter 7).

d. Some people might criticize P, arguing that blocks should have been included in his original analysis. To be consistent, this argument would have to utterly reject any study that failed to include blocks in the design; such designs cannot even do the indicated analysis. In contrast, P's design not only balances the block effect over treatments, but provides a measure of size and reliability of the block main effect.

I would not criticize P in this way. I suggest that including blocks in a complete Anova would be the poorer choice, failing to give adequate weight to extrastatistical background knowledge. Following the rationale of partial analysis of Section 5.3.4, however, the main effect of blocks deserves scrutiny.

a5. Design an experiment in your area in which you would use the initial phase to increase reliability and validity of the response measure. Specify exactly how you would expect each advantage to be realized.

a6. In the field study of smoking prevention cited in the Appendix to Chapter 3, the effect at 12 months was not large.

    a. How might the screening procedure be amplified to screen out more of the probable relapsers?

    b. What advantage might be expected from (a)?

a6ANS. Presumably the relapse rate is largest early in the quit attempt. If these early relapsers could be screened out, the social resources could be used more effectively on those more likely to quit. One way to improve screening effectiveness would be to ask all persons to wait a week for a callback to schedule them in the program. Those weak in quit potential will be more to have lost interest after a week has passed.

One advantage of an improved screening procedure would be higher success rate.

a7. The last paragraph under *Screening Tests* states that limiting the subject population can increase process validity. Cite two different situations in which you could expect such increases.

a7ANS. Process validity can be increased by screening out subjects weak on the process. In studies of color vision, for example, those weak in color vision are commonly screened out. In a few studies, however, those weak in color vision may be screened in to study the nature of this defect. Screening is often used in language studies for similar reasons. A related example is the screening of the very poor subjects in the anagram experiment of Section 14.2.2. A rather different reason for screening is to avoid confounding the process of interest with some other process. Thus, a study of second language learning would no doubt screen on language background.

a8. a. Find a published experiment in your area in which you would expect a benefit from blocking on individual differences in follow-up work. Specify the blocking variable, how you would measure it, why you chose the number of blocks you did.

b.* Include a hopefully realistic guess of how much reduction in MS_{error} you would expect; (Equation 15 from Chapter 9).

a8ANS. b.* In Equation 15 of Chapter 9, $r^2 = SS_{pred}/SS_Y$, where SS_Y corresponds to the total error variance in Y, and SS_{pred} corresponds to the amount of SS_Y predictable from the individual difference variable. Hence r^2 is the proportional reduction.

In this way, the question reduces to a guess about r, which may be done in a framework of background information about correlations of individual differences in general and in the particular task at hand.

a9. How could you use block design to deal with the threat that knowledge of a deception might spread through the subject population in the third example of Note 8.1.5a?

a9ANS. If knowledge of the deception does spread and if this influences subjects' reactions, this will appear as a trend across blocks. In the worst case, the later blocks could be eliminated. The danger seems small, especially compared to the threat of confounding of treatments with time in this particular experimental situation.

b1. In Table 14.4: a. Verify the numerical values for each of the listed SSs.

b. Show that F_A would not be statsig had the position effect been ignored.

b2. Q tested five treatments in a Latin square, with two groups of five subjects, one experimental and one control.

a. Write out Source and df for her Anova table, ignoring the Latin square and treating this as an $(S \times A) \times G$, repeated measurements design.

b. Extend (a) to a Latin square analysis by including the position effect. Indicate what each Source means.

b2ANS. a. This Anova table has the very same format as Table 6.3. In the Between subtable, F_G has 1/8 df. In the Within subtable, F_A and F_{AG} both have 4/32 df.

b. Calculate SS_P on 4 df by hand as in the text. This SS_P is included in $SS_{[SA]}$, from which it should be subtracted, the residual being the error.

b3. Section 14.3.1 suggests beginning the experiment with the same practice session for all subjects in order to reduce the position effects during the experiment proper. Cite a specific task in which this procedure would be (a) inadvisable (b) advisable.

b4. The second paragraph under *Startup Effects* states that the preliminary phase has a "threefold purpose" (page 426). Describe a specific experimental task in your own field to illustrate how each purpose would be achieved.

b5. You use a 5×5 square but with two successive replications for each subject. You have at least three different choices for order of treatments in the second 'replication. What are their advantages and disadvantages?

b5ANS. The order of treatments in the second replication could be the same as in the first, in the reverse order, or in a new random order. Position effects are handled equally well by all three choices. But carryover effects are balanced better with the second and third choices. I incline toward the second choice, using a balanced square.

b6. You give four treatments in succession to each subject. Unknown to you, all have equal effect. The response changes, however, because there is a position effect. Suppose this position effect is entirely linear.

a. Show by numerical example (or better yet, algebraically) that giving the treatments in the forward and reverse orders will balance out this linear trend.

b. Does similar balance hold for any number of serial positions?

c. Suppose the treatments have different effects. Do the forward and reverse orders still balance the linear position effect?

b6ANS. c. Suppose the linear trend increases response by c at each successive serial position. The mean response to A_j becomes $\bar{Y}_j + \frac{1}{2}(a-1)c$. The $\frac{1}{2}(a-1)c$ cancels in comparing differences among treatment means.

b7. You run four subjects in a repeated measures design with four treatments. Although there is good reason to believe position effects are negligible, you balance them with a Latin square. Now you wonder whether to use the repeated measures analysis of Chapter 6 or the Latin square analysis of this chapter.

a. How many df does the Latin square give you for error?

b. What is the advantage of including the position effect in this Anova?

c. What is the disadvantage of including the position effect in this Anova?

d. If position is not included in the Anova, is the position balance destroyed?

e. You consider increasing the error df by ignoring position and treating this as a simple repeated measures design. What do you decide?

b7ANS. a. Table 14.5 shows 6 df for error, with $N = 4$ for the single square.

b. By including position, confidence intervals can be obtained for position effects, an indication of their magnitude.

c. The disadvantage of including position is that it uses up 3 df in the Anova. These 3 df could be included with error on the ground that position effects are negligible. With a single 4×4 square this would increase error df from 6 to 9, which could give substantially narrower confidence intervals for treatment means.

d. Position is balanced by virtue of the Latin square design, regardless of the Anova or any other statistical analysis.

e. With so few df for error, I might leave the position effect in with the error. A large position effect would inflate the error and reduce power. But since the position effect is expected to be small, the gain in df for this small square should overweigh any bias from including it in the error (Section 18.4.4).

The best action, of course, is to replicate the square, which would increase error df and still include position as a source. In an empirical view, so small an experiment should usually be expected to present uncertainty about the most appropriate analysis and to leave considerable uncertainty about the results.

b8. You wish to balance position effects in your experiment, in which each subject gets eight treatments. Unfortunately, apparatus limitations allow you to use only four sequences. How can you apply the Latin square idea to get partial position balance?

b8ANS. Pairing treatments, indicated by enclosure in imaginary parentheses, yields a 4 × 4 Latin square that provides partial position balance and $SS_{position}$ on 3 df.

$$(12)\ (34)\ (56)\ (78)$$
$$(78)\ (56)\ (34)\ (12)$$
$$(56)\ (78)\ (12)\ (34)$$
$$(34)\ (12)\ (78)\ (56)$$

b9. You have a 2 × 3, repeated measures design in which each subject takes all 6 treatments. To assess position effects, you use a 6 × 6 Latin square to balance treatments across position. Write out Source and df for your Anova.

b9ANS. The Anova table is the same as given at the left of Table 14.5, except that the A effect is broken down into three parts: The main effects of the two variables plus their interaction residual, on 1, 2, and 2 df, respectively. These may be calculated as shown in the Appendix to Chapter 5.

b10. You have three infant monkeys in each of Harlow's three mother love conditions, each infant to be tested in all of three social interaction situations, using a Latin square.

a. Write out Source and df for the Anova table, ignoring the Latin square and treating this as an $(S \times A) \times G$, repeated measures design.

b. Extend (a) to a Latin square analysis by including the position effect. Indicate the error term for each source and what it would mean empirically.

b10ANS. a. The Anova table follows the format of Table 6.3 exactly. The Between subtable has 2 df for mother, 6 df for error. The Within subtable has 2 df for social situations, 4 df for mother × social situation, and 12 df for error.

b. To extend the repeated measures analysis of (a) to include the position effect, follow the schema of the right half of Table 14.5, as specified under *Including Between Subject Variables*. The Between subtable remains as in (a). The SS for position on 2 df is calculated as in Table 14.4 and subtracted from the interaction, $SS_{[AG]}$, in the Within subtable as indicated in Table 14.5.

b11. The 2 × 2 Latin square presents unique difficulties in interpretation. As one example, the position effect in a 2 × 2 Latin square is sometimes interpreted as an interaction between treatment and sequence. In an empirical study of drug effects on memory cited by Cotton (1989; see further Dawes, 1969; Hills & Armitage, 1979), placebo means were approximately 3.4 in the placebo–drug sequence and 4.6 in the drug–placebo sequence; drug means were 3.2 and 2.0 for the placebo–drug and drug–placebo sequences, respectively.

a. Plot these data as a 2 × 2 factorial graph, with drug and placebo on the horizontal axis, one curve for the placebo–drug sequence, and one curve for the drug–placebo sequence. Does this graph support a treatment × sequence interaction? At face value, what does this mean about

transfer?

b. Replot these four means as a 2 × 2 factorial graph with drug and placebo on the horizontal axis, one curve for each position. Reinterpret by visual inspection.

c. Which of these two graphs seems to give a truer picture of the data? Or do both have equal validity? Justify your choice.

d. Suppose the placebo mean in the drug–placebo sequence was 4.2 instead of 4.6. How would this appear in the Anova of (b)? What would this mean for interpreting the data?

b11ANS. a. This graph shows a crossover, a treatment × sequence interaction. At face value, this crossover implies that giving the drug first produces a large transfer to the placebo; but giving the placebo first produces little transfer to the drug. Most important, the placebo seems to facilitate the drug effect.

b. In this format, the two curves are parallel. Drug and placebo both yield better memory when they appear second. This format exhibits a strong position effect.

c. I consider that the second graph gives a truer picture; practice effects may be expected in this memory task. This opinion is based on extrastatistical background knowledge that the position effect in (b) has plausible psychological explanation, whereas the treatment × sequence interaction in (a) does not.

d. The factorial graph will now be nonparallel—an interaction. But this interaction requires little qualification of main effects, in line with the theme of Chapter 7.

Statistically, position and differential carryover effects are completely confounded in the 2 × 2 design. Interpretation depends on empirical knowledge.

b12. Consider the following data for the Greco–Latin square of Table 14.6.

$$\begin{array}{cccc} 4 & 8 & 11 & 11 \\ 7 & 9 & 12 & 10 \\ 10 & 10 & 9 & 13 \\ 13 & 11 & 10 & 12 \end{array}$$

a. Show that SSs for rows, columns, A, and B all equal 20.

b.* By what simple rule were these data generated?

b13. Assume treatment effects of 1, 2, 3, and 4 for A_1, A_2, A_3, and A_4, respectively, in a Latin square design. Also assume that A_4 has a carryover effect of 4, which is added to whatever treatment follows. Assume all other effects are zero in the Latin square model of Section 14.3.4.   a. Use the model to generate the 16 cell entries for the cyclic square of Table 14.3 and for the balanced square of Table 14.7.   b. Get treatment means for both squares; explain the difference.   c. Get position means for both squares and explain.

b13ANS. a. Cell entries for the cyclic and balanced squares are:

$$\begin{array}{cccc} 1 & 2 & 3 & 4 \\ 2 & 3 & 4 & 5 \\ 3 & 4 & 5 & 2 \\ 4 & 5 & 2 & 3 \end{array} \qquad \begin{array}{cccc} 1 & 2 & 3 & 4 \\ 4 & 7 & 2 & 1 \\ 2 & 4 & 5 & 3 \\ 3 & 1 & 4 & 6. \end{array}$$

c. The position effect is a confounded carryover, not genuine.

b14. A single subject is presented treatments using an $a \times a$ Latin square, so a responses are obtained for each treatment. The subject goes through the successive treatments in the first row, then in the second row, and so on.

a. You apply the Anova outlined in the left part of Table 14.5. What does Subjects in the Between subtable correspond to in this design? What does this term mean? How does this term relate to the position effect?

b. How would you test the treatment effect? Is the ε adjustment needed?

c. What would you do if only some rows of the square could be run?

b14ANS. a. The Subjects term in Table 14.5 corresponds to the a successive rows in the Latin square. Differences between rows could arise if, for example, practice effects were mainly in the initial row.

b. For statistical analysis, responses should be independent. The Latin square imposes a partial treatment randomization that promotes independence but does not guarantee it. It seems desirable, therefore, to minimize carryover effects experimentally.

On this basis, the error term given by Table 14.5 would be used for the treatment effect and for the position effect as well. This error term is quite different in nature from the SA interaction in the regular Latin square, since it represents variability in the responses of the single subject. No ε adjustment is needed; this is not a repeated measurement design with correlated scores.

c. Treat the design as a factorial with r rows and a treatments. I would choose rows with some kind of forward–reverse order balance (Exercise b8).

b15. For each of the following situations, give your opinion whether a between subjects or within subjects design should be used, with justification.

1. Harlow's study of wire–cloth–real mothers.

2. Line–box illusion of Figure 18.1.

3. Effects of lesions in lateral and ventromedial thalamus (Section 8.1.6).

4. Methods of teaching psychology in high school.

5. Prototype experiment of Section 8.1.2.

6. Blame experiment of Figure 1.3.

7. Study of effects of motivation on perception of coin size in Section 8.1.2.

8. Rats tested with varied amounts of reward on one variable interval schedule.

b15ANS. Between subject design cannot be avoided for Harlow's mother love situation, cannot generally be avoided in studying methods of teaching in high school, and would usually produce undesirable interference with thalamic lesions. Rats will take some time to adjust to each new amount of reward; this adjustment period will show carryover from the previous reward as well as position effects until stable state develops. If interest is only in stable state response, it might be reasonable to expect no carryover effects. For the other five situations, within subject design would seem usually feasible and desirable.

b16. This exercise aims to bring out potential for adapting Latin squares to new situations. In the study of memory organization in the last subsection of Section 8.1.6, write out a Latin square you might use that would avoid the confounding and yet still allow within subject design for comparing how much each of the four types of transformation degrades memory. Assume only four different original scenes were used.

a. Write out source and df. Say what each source would mean.

b. What would your response measure be for each scene?

c. What would your response measure be for the Anova?

d. What additional variable needs balancing in presenting treatment conditions?

b16ANS. The following design is one of many possible. For the four rows of the square, use the orders {1, 2, 3, 4}, (2, 1, 4, 3}, {3, 4, 1, 2}, and {4, 3, 2, 1}. Almost any square would do the main job, but this one looks good for the balance of (d). The first treatment in the square thus represents two scores: one original scene plus one of its transformations. Each successive treatment represents a different original scene plus a different one of the four transformations. Thus, each row of the square requires eight responses.

I would give the two copies of each scene in original–transformed order for the first two treatments for the first two subjects in the square, in the transformed–original order for the last two treatments for these same two subjects. For the second two subjects, the opposite orders would be used. This balances the original–transformed order and allows it to be measured.

Seeing two similar scenes in succession may perplex subjects. Perhaps the best solution is to tell them about this beforehand, and/or to include some additional original scenes as initial practice in the memory session.

I would use a numerical judgment of familiarity or of recognition confidence for each scene. Numerical judgment is more sensitive than choice response. For Anova, I would use the difference in judgment for each original and transformed scene.

I would make up hypothetical data for a dry run analysis to make sure this Anova would work.

ANSWERS FOR CHAPTER 15

Exercises 1–7 and 19 pertain to the nursing home study of Section 15.5.2.

1. List three alternative explanations for the claimed improvement of the experimental group in the nursing home experiment of Section 15.5.2.

1ANS. One alternative is that the experimental group did not really improve. What was statsig was the difference between experimental and control, but the authors remark that the control apparently deteriorated substantially, as noted in the text (page 467). The result could be due to this control deterioration alone.

   An entire class of alternatives involves possible differences in happenings on the two floors of the home—after the treatment but before the measurement—that could have caused the difference even though the treatment itself had no effect. Examples include a new nursing attendant, deaths of one or two prominent residents, and so forth. This is the *History* confound discussed in Section 15.5.3 on Quasi-Experimental Design.

2. Suppose the experiment had been replicated using the first and third floors in the same way as was done for the second and fourth floors. What advantage would accrue?

2ANS. Floor is a natural group in this study (Section 15.1) Replication with the two other floors would provide a measure of differences between floors treated alike. The actual design confounds floor differences with treatment conditions. If this replication showed floor differences to be small, relative to the difference between treatments, substantially more faith could be placed in the claim for a treatment effect.

3. An alternative method would be to assign the two treatments at random to individual residents within each separate floor. What are the advantages/disadvantages of within floor randomization?

3ANS. The most important advantage of within floor randomization would be to avoid the confounding of floor with treatment, a grave threat considered in the two previous exercises. In addition, it could avoid the confounding of treatments with the views of the home administrator about the effectiveness of those treatments, as by presenting written versions of the treatments separately to each resident.

   Randomization within floors requires more time to present the treatments to each resident individually. Possibly, some adverse reaction may arise if some residents learn that they are getting one treatment while others are getting another.

4. An alternative design for the nursing home study would have given no specific talk or other treatment to second floor residents, but still make the same measurements as for the fourth floor residents. However, the control condition actually used was needed to control a confound in the experimental condition given the fourth floor residents that would affect the theoretical interpretation. What was this confound?

4ANS. In this nursing home study, the causal process at issue was personal control. The primary experimental treatment was intended to increase residents' feelings of personal control; this was expected to increase feelings of well-being. But this treatment also gave considerable personal attention to the residents, which might have similar effects (as in the well-known Hawthorne study). This personal attention was thus confounded with the hypothesized process of personal control. The second floor treatment deconfounds personal attention and related factors other than personal control.

5. The main data were change scores between measures for each resident before and after the treatment.
   a. What is the advantage of this change score?
   b. What is the disadvantage of this change score (see Chapter 13)?

5ANS. a. The apparent advantage is that the change score seems to control for main effect of individual differences.

b. The disadvantage is that the change score is less powerful than analysis of covariance (Section 13.1.2). Indeed, this change score could be more variable and so less powerful than the after score alone.

6. "Locus of control" is one of the few aspects of personality that has shown substantial correlations with performance in a variety of tasks. Locus of control refers to the degree to which people feel that they control external events or that external events control them. How might a preliminary measure of locus of control have been used in the nursing home study to:
   a. Decrease error variability? How much decrease would you expect?
   b. Increase substantive significance?

6ANS. a. The measure of locus of control could be used as a blocking variable (Section 14.2) or as a covariate in analysis of covariance (Chapter 13). Locus of control seems rather similar to "personal control," so at a guess it might correlate between .4 and .5 with the measures of well-being in the nursing home experiment. In that case, it would reduce error variance about 20%, well worth the small cost of measuring locus of control.

b. Locus of control would seem to be closely related to the concept of personal control used by the investigators, so its inclusion could help provide construct validity for their concept of personal control. The extensive literature on locus of control would become relevant. In addition, the results might show that the treatment effect differed for persons with different locus of control, perhaps being most effective for persons with medium scores. This would be a step toward improving the treatment.

7. For a follow-up design for the nursing home study, assuming comparable facilities are available, what two changes would you consider most important?

7ANS. I consider within floor randomization essential. Within floor randomization can avoid most confoundings noted in previous exercises, as well as others not mentioned. Without randomization, the study seems questionably worthwhile.

Except for the plant, the experimental manipulation was not maintained over time, and so unlikely to have enduring effects (see Exercise 10.6). To maintain effectiveness, it may be necessary to develop a treatment run largely by the residents themselves, with modest supervision by the staff or by community volunteers. One possibility would be to encourage small interest groups of residents, on exercise, reading, or talking. Autobiographical groups, in which each resident tells their life story in installments, might be viable (see Rokeach, *The Three Christs of Ypsilanti*, 1964).

8. This exercise aims to make concrete some issues in random factor analysis. The following experimental situation is typical of one class of situations in which random factor analysis has sometimes been misapplied.

Subjects' responses may depend on personality characteristics of some person that are confounded with the treatment. Studies of social influence, for example, may use a confederate who has some essential role in the treatment. Similarly, effectiveness of marriage counseling will depend on the personality of the counselor. Again, expression of emotion in animals may depend on how the experimenter handles them.

To handle such confounding, it is often vital to use more than one confederate, counselor, or experimenter. For illustration, we may consider the study of social interaction discussed in Exercise 5.10, simplified for present purposes by ignoring the Favorable-Unfavorable manipulation. Suppose two confederates were used, together with an experimental variable in a 2×2, Confederate $\times$ Attractive-Unattractive design, with $n = 12$.

a. What is F^* if confederate is treated as a fixed/random factor?

b. All MSs are the same, regardless of whether confederate is treated as random or fixed. What is changed in the Anova table?

c. What are two undesirable consequences of treating confederate as random?

d. Suppose F was statsig with confederate treated as a random factor. To what population do you think this result can be generalized?

e. If confederate is treated as fixed, what can be said about generality?

8ANS. a. $F^*(1, 44) = 4.06$. $F^*(1, 1) = 161$.

b. What changes in the Anova table is the MS used for error, and this depends on whether factors are considered fixed or random.

c. Statistically, two consequences of treating confederate as random are to increase the error term and to increase the criterial F^* for the main effect of the experimental variable. Power suffers from both consequences.

d. In principle, the result can be generalized to the population from which the two confederates were taken—if they were taken randomly. But the large $F^*(1, 1) = 161$ suggests this test would have too little power to have any value as evidence.

Furthermore, the population from which the random sample is drawn will usually be a handy population, small and poorly defined. Hence the generality of the result still depends on extrastatistical inference.

Similar results with two confederates is of course substantially more reassuring than a single confederate, but this reassurance is largely extrastatistical.

e. Generality depends primarily on extrastatistical inference, regardless of whether confederate is treated as fixed or random. Of first importance is the main effect of confederate and whether treatment effects show the same pattern for both confederates. If confederate main effect is small, and/or if treatments show the same pattern for both, some mild confidence about generality seems in order. Two confederates is a narrow base for generality, but much better than one.

9. According to Section 15.1.2, individuals from the same natural or ad hoc group are likely to be more similar than individuals from different groups.

a. How does this statement manifest itself in MS terms?

b. How can you test this statement statistically?

9ANS. a. The *expected* MS for groups is larger than for subjects within groups. Because subjects are more similar within groups, the groups themselves differ more than the subjects within them.

b. By Table 15.1, test $F_{groups} = MS_{G/A}/MS_{S/G}$.

10. Fractional replication design can be useful for studying long serial curves. To illustrate with a memory issue, consider judgments of frequency of occurrence, which some writers have claimed involves automatic processing unaffected by practice and experience. To measure the serial curve of frequency judgments, suppose a subject sees a rapid sequence of letters, followed by a single letter. The subject's task is to rate on a graphic scale the relative frequency of the single letter in the sequence.

In principle, a serial-factor design can be used, in which each serial position represents a factor with 2 levels, namely, presence or absence of the given single letter at that serial position. The main effects of this design measure the relative effect of each position on the response at the end.

In practice, fractional design seems essential. For 12 serial positions, which seems pretty minimal, a complete factorial design would yield $2^{12} = 4096$ sequences. With fractional replication, 16 sequences would suffice.

Further reduction in design size can be gained by coupling serial positions, so that each factor in the design is a block of successive serial positions, still with two levels, according as there are, say, 0 or 1 instances of the given single letter in that block.

To illustrate with a too-simple example, show how the fractional design in Table 15.3 could be used to get the serial curve of frequency memory for a single subject for 12-letter sequences, coupled in blocks of 4, with only 4 trials.

10ANS. The factors of the 2^3 design are the three successive blocks of four trials. Each block has, say, 0 or 1 occurrences of a specified letter. Hence each factor has the same frequency value; its main effect is the memory effect, so the serial curve of main effects is the serial curve of memory. This example is unrealistically small, of course, but the idea can be used for much larger sequences with smaller blocks. Some comments on serial memory curves are given in Anderson (1996a, p. 388, Note 5).

11. In Section 15.5.3, the subsection *Differential Growth* cites two reasons to expect higher growth rate for the C group in Head Start programs. Draw graphs to illustrate how each reason will cause an experimental treatment that has no effect to appear harmful.

11ANS. The first reason, higher growth rate in previous years, implies a curve with greater elevation and greater slope for group C, not merely before, but also during the time period of the study. For simplicity, take this curve to be a straight line. Since this line has greater slope for group C, its change from beginning to end will be greater; two vertical lines on the graph may be used to exhibit the E − C difference before and after. The after difference will be greater, implying that the treatment was harmful.

The second reason, differential growth, may be illustrated with the same graph. In this case, however, the greater slope for group E is a consequence of their greater elevation at the beginning. Of course, both processes may act jointly.

12. For *Differential Growth* in Section 15.5.3, draw a graph to show that if an ineffective treatment is given to the initially superior group, it will falsely appear to have positive benefits (Section 13.2).

12ANS. Just switch the E and C labels on the graph from the previous exercise.

13. Under *Error df* in Section 15.1.2, justify the statement that the expected value of the pooled error is generally smaller than the correct error term.

13ANS. The correct error term is $MS_{G/A}$ from Table 15.1. $MS_{S/G}$ is too small because its expected value does not include the term in $\sigma^2_{G/A}$. Pooling means adding $SS_{G/A}$ and $SS_{S/G}$ and dividing by the sum of their df. In this weighted average, the too-small error term usually gets by far the larger weight. Although there may be circumstances in which pooling is justified, clear and explicit justification is needed (Section 18.4.4).

14. In the example of the half replicate of a 2^3 design considered in Table 15.3:

    a. Guess what is confounded with main effect of B; of C.

    b. Prove your guess using the ± signs in the table.

14ANS. b. The boldface ± signs for B are +, −, +, −, in order. Those for AC are exactly the same, so the two are completely confounded. Similarly, the boldface ± signs for both C and AB are +, −, −, +, in order.

15. This question asks for *normalized* confidence intervals for fractional replication with independent scores. To normalize, divide the positive c_j by a constant so they sum to 1; do the same for the negative c_j; then the contrast has the same unit as the data (Section 18.2.1, page 560). Use Expression 4 to show that normalized confidence intervals are:

$\hat{\kappa} \pm t^* \sqrt{MS_{error}/2n}$, for a half replicate of a 2^4 design.

$\hat{\kappa} \pm t^* \sqrt{MS_{error}/n}$, for a quarter replicate of a 2^4 design.

$\hat{\kappa} \pm \frac{1}{2} t^* \sqrt{MS_{error}/n}$, for a half replicate of a 2^5 design.

16. Show that the confidence intervals given in Expressions 6 and 7 reduce to those given in Chapter 5 when all *n*s are equal.

17. For the Latin square in Section 15.3.2, verify that "Each level of any one variable occurs exactly once with each level of each other variable." List the index numbers for the 16 treatments in a three column array, headed by A, B, and C. Then it is easy to check that each combination of two index numbers occurs once and only once.

18. In the $(S \times A) \times G$ design, is A nested in or crossed with G?

19. In the nursing home study, what confound in the control condition actually used might undercut the theoretical interpretation in terms of increased feeling of personal control by individuals in the experimental group?

19ANS. The second floor treatment emphasized the responsibility of the staff to care for the residents; this actively de-emphasized personal control of the residents. A difference between floors might thus be due to a loss of feeling of personal control by the second floor residents, not a gain of feeling of personal control by the fourth floor residents.

ANSWERS FOR CHAPTER 16

1. In the Oregon admissions procedure, it might be objected that this prediction equation would be unfair to applicants from minority groups. How do you suppose they handled this matter?

1ANS. Minority applicants were excepted from admission by the formula and evaluated case by case.

2. In relation to *Missing Variables Confounding*:

   a. Why might height have a large *b* weight for predicting childrens' vocabulary size if age was missing?

   b. Do you think one is a better predictor than the other?

   c. Do you think one is more causal than the other?

   d.* Can effects of age and height be separated?

2ANS. a. Height has a high correlation with age and hence a high correlation with vocabulary size. If age was missing, height could act as a surrogate measure of age.

b. Age will no doubt be mildly better than height because it is more highly correlated with grade in school.

c. Height is surely not causal. Still, age is not causal in itself; vocabulary size is caused by processes that occur in time, not by passage of time itself.

d.* Statistics cannot separate effects of correlated variables without reliance on extrastatistical information. In this artificial example, extrastatistical information argues against including height in the regression. In real applications, extrastatistical information is not usually this clear.

3. In *Mystery of the Missing Cloud Cover* in Section 16.2.2, explain:

   a. How heavier fighter opposition could cause lower bombing accuracy.

   b. How the cloud cover variable could produce the counterintuitive result.

3ANS. a. Heavier fighter opposition could cause the bomber pilot to take evasive action, increase speed, and increase altitude, all of which would decrease bombing accuracy.

b. Cloud cover has a substantial negative correlation with enemy fighter opposition. Lacking good radar at that time, fighters could not locate bombers in heavy cloud cover. Bomber losses must have increased without clouds, but those that got to their targets had greater accuracy. Enemy fighter opposition thus served as a surrogate measure of cloud cover, just as height for age in the previous exercise.

4. *Mystery of the Missing Cloud Cover* of Section 16.2.2 shows that a missing variable can have fatal consequences. Experimental studies, of course, typically manipulate only a few variables. Numerous variables are thus missing from the design.

   a. In what way are randomized experiments not troubled by missing variables?

   b. In what ways are randomized experiments troubled by missing variables? Cite specific experimental illustrations from previous chapters.

4ANS. a. With randomized experiments, the manipulated variables produce the observed effect. No observed effect can be due to a missing variable (barring a "bad" randomization) because the variable does not vary systematically across experimental conditions and so is uncorrelated with conditions.

b. Two general limitations on randomized experiments arise from missing variables. First, a missing variable may qualify the action of a manipulated variable. This occurs with the crossover interaction in the left panel of Figure 5.4. Had only variable *A* been manipulated, it would necessarily have yielded a misleading result. A real example appeared in the praise–blame experiment of Note 5.2.2a, in which the reinforcement effect would have been considered small had not the personality variable been included. Other crossovers are noted in Section 14.4.

Markedly more common is confounding: A manipulated variable may have two distinct effects. The classic example is medical treatments, which may have true medicinal value together with a placebo effect.

In general, confounding means that the missing variable problem is serious for process analysis based on randomized experiments in the same way as for multiple regression of observational data (Chapter 8 gives numerous examples).

Randomized experiments have priceless advantages, therefore, because they can sometimes avoid known confounds and can sometimes allow experimental control of confounds. Indeed, knowledge about confounds and how to control them experimentally is a major component of research judgment. Such knowledge is basic for planning good experiments.

5. Under *Cognitive Analysis of Clinical Judgment* in Section 16.1.3:

   a. Justify the statement that "If subjects did make configural judgments, the additive model should fail to fit them."

   b.* Make up a factorial design using trait adjectives to describe persons and select row and column adjectives that you think will produce configural response, that is, responses that depart in some systematic direction from additivity. Make up and plot hypothetical data.

5ANS. a. If the subject's judgments are configural, then the meaning and value of an adjective will vary from one cell to another. The additive model, however, postulates constant values. Hence it cannot be correct if configurality holds.

b. Perhaps the surest way to get configural response is to use inconsistent adjectives, such as *honest* and *dishonest*. Such inconsistency can be made realistic by attributing each adjective to a different acquaintance of the person. The resolution of this inconsistency can be controlled by including a third adjective of positive or negative value. An alternative is to use redundant adjectives attributed to the same acquaintance.

A reference standard is also needed. This can be obtained by including the given description as part of a factorial design, thus allowing comparison with other cells in which no configurality is expected. Inconsistency resolution will then appear as a directionally predictable deviation from parallelism (see Anderson, 1981, Section 3.4.1).

6. In *Taxes and Education* (Section 16.2.2):

   a. Give two plausible interpretations of the statsig *b* weight for background variables.

   b. What relevance does the statsig *b* weight for background variables have to the cited interpretation of the nonstatsig *b* weight for resource variables?

6ANS. a. Two plausible interpretations of the influence of background variables are environment and heredity. Parents with higher education level are more likely to have a more intellectual home environment, with books, conversation, and so forth, that will exert a positive effect on their children's grades. Also, parents with higher education level tend to have higher intelligence, which will tend to be inherited by their children.

b. The statsig weight for background variables shows that the regression analysis has adequate power to detect real effects. The *N* of 88 is not especially large for regression analysis and so comes under question with respect to power.

7. One of the three studies on effects of beta carotene (vitamin A) on cancer cited under *Personal Health* in Section 16.2.2 found a statsig harmful effect. The article discussed this finding and concluded "In spite of its formal statistical significance, therefore, this finding may well be due to chance." Is this a proper way to gloss over a statsig finding?

7ANS. The results of any experiment should be interpreted in light of other relevant evidence. False alarms do occur, but some of them can be detected because they contradict other knowledge. In this case, the article noted that none of the few other studies had found a harmful effect and that "There are no known or described mechanisms of toxic effects of beta carotene, no data from animal studies suggesting beta carotene toxicity, and no evidence of serious toxic effects of this substance in humans" (*New England Journal of Medicine*, 1994, *330*, 1029-1035). On the basis of evidence then available, their interpretation seems appropriate. Surprisingly, the harmful effect may be real (see Note 16.2.2c).

8. a. How does the sex–marital satisfaction example on page 503 relate to the subsection on *Person–Variable Confounding*?

   b. Hypothesize a similar example in an area of interest to you.

8ANS. a. The variables are amount of intercourse and marital satisfaction. Couples who do have more sex are more satisfied. To suggest that less satisfied couples would be happier if they had more sex overlooks that they are different people from those who do have more sex. The confounding is between two general causes of the observed effect.

b. Person–variable confounding is the essence of many, many applications of regression analyses. Most evaluations of Head Start programs are of this type, as are the effects of low blood lead on children (Section 16.2.2). Most reports in the media about effects of different foods on health and disease involve similar confounding. They may be correct, but better evidence is needed before putting much faith in them (see *Personal Health* in Section 16.2.2 and Note 16.2.2c).

9. Under *Clinical Versus Statistical Prediction 1* in Section 16.1.3, Sarbin's test of clinicians' skill suggested that clinical judgments added little to the regression prediction.

   a. Outline the steps needed to demonstrate this conclusion.

   b. Is this a fair test of clinical expertise?

9ANS. a. Sarbin's analysis employed the following three steps.

1. Run the two-variable regression using the two cited predictors (X_1 = percentile rank in high school, X_2 = score on college aptitude test).

2. Run a three-variable regression, adding X_3 = grade predicted by the counselor.

3. Test whether X_3 adds real predictive power as indicated under *Assessing Single Variables* in Section 16.1.1.

b. It seems a fair test whether the counselors could predict students' grades better than the two cited scores. Since grades depend heavily on motivational factors and work habits, and since the counselors had much more relevant information available, it is surprising that they did no better. They did, however, exhibit fair predictive ability, at least with female students.

10. Show how the issue of gender bias in the last subsection of Chapter 10 can be conceptualized in terms of correlation produced by a missing variable.

10ANS. The 2×2 contingency table indicated in the first paragraph of *Missing Variables* at the end of Chapter 10 reveals a correlation between gender and acceptance in graduate school. This correlation vanishes if the missing variable of department (acceptance rate) is added.

11. Aptitude–treatment interactions have been a central concern in educational psychology on the plausible ground that the most effective procedure for teaching depends on the aptitude of the learner. Let *X* denote the aptitude, *Y* the amount learned. Consider an experiment with two treatments (e.g., *Methods for Language Learning* in Section 14.2.2).

   a. How will an aptitude–treatment interaction appear graphically?

   b. How will an aptitude–treatment interaction appear in one-variable regression analysis?

   c. Referring to a comparison of regression lines between treatment conditions, Pedhazur and Schmelkin (1991, p. 547) say that "In short, *a conclusion that there is no significant difference between the b's* [of the one-variable regressions] *is tantamount to a statement that there is no interaction between the treatments and the attribute* [attribute refers to aptitude of subject]." Comment.

11ANS. a. Graphically, aptitude–treatment interaction will appear as different slopes of the plots of *Y* versus *X* for the two treatments.

b. The different slopes in (a) will appear as different *b* weights in the one-variable regressions for the treatments.

c. The quotation embraces the null hypothesis of no interaction. More useful would be a confidence interval for the difference between two *b*s. Elsewhere, Pedhazur and Schmelkin give careful warnings about accepting a null hypothesis, but in the discussion surrounding this quote (their italics) they are quite set on doing so.

12. Suppose you are using a regression with unit weighting, these weights being given. How can you test whether some one of the given variables can be eliminated without undue loss?

12ANS. Run two regressions, one with and one without the given variable. The difference between the two in SS_{pred} measures the loss from elimination of the given variable. A significance test can be made as in *Assessing Single Variables* in Section 16.1.1.

13. "A variable that is a good predictor by itself may add little when assessed as one of a group." (From *Relative Importance* in Section 16.1.1.)

a. Verify this quote for the extreme case of a three-variable regression based on X_1, X_2, and $X_3 = X_1 + X_2$ (all intercorrelations positive).

b. Based on (a), why does this quote apply generally?

c. How did this issue apply to the prediction equation for success in graduate school?

13ANS. a. The variable $X_3 = X_1 + X_2$ will add no predictive power because its predictive power has already been used up in the two-variable regression. (If X_3 is measured independently, it will increase predictor reliability and thereby increase predictive power although probably not much.)

b. Because X_3 will generally be correlated with X_1 and X_2.

c. The prediction equation for success in graduate school originally included the third variable of quality of undergraduate school. This was a reasonably good predictor of itself, but it added little to GRE and GPA (page 492).

14. Write down two-variable linear regression equations for the following four sets of four data points using graphs and visual inspection only. Comment.

a. X_1	X_2	Y		b. X_1	X_2	Y
1	2	1		1	2	5
2	2	2		2	2	6
3	2	3		3	2	7
4	2	4		4	2	8

c. X_1	X_2	Y		d. X_1	X_2	Y
1	1	1		1	5	1
2	2	2		2	6	2
3	2	3		3	7	3
4	1	4		4	8	4

14ANS. a. $Y = X_1$. b. $Y = X_1 + 4$. c. $Y = X_1$.
d. $Y = \frac{1}{2}X_1 + \frac{1}{2}X_2 - 2$. In (a) and (b), X_2 has zero range and hence no predictive power. Of course, X_2 may have a strong relation to Y that is masked by its lack of variation. In (c), X_2 is quadratic and uncorrelated with Y, which is linear. In (d), X_1 and X_2 are perfectly correlated so many other regressions are possible, such as $Y = X_2 - 4$.

15. The discussion of nonlinearities in Section 16.1.5 implies that artifactual nonlinearity can also appear if any X_j is measured on a nonlinear scale. Demonstrate this by considering the case

of $Y = \psi_1 + \psi_2$, where ψ_1 and ψ_2 are linear scales of the concepts that X_1 and X_2 are supposed to measure. Assume that X_1 is a linear scale, with $X_1 = \psi_1$, but that X_2 is a nonlinear scale, with $X_2 = \sqrt{\psi_2}$. For simplicity, assume also that X_1 and X_2 are uncorrelated.

15ANS. By hypothesis, $Y = \psi_1 + \psi_2$. This relation becomes $Y = X_1 + X_2^2$, by the given relations between X_j and ψ_j. This last relation is in terms of the observables, and it is nonlinear in X_2. Without knowledge of the true scales, there is no sure way to rule out this possibility and hence no sure way to interpret nonlinearities in terms of process or genuine interaction. This obstacle is serious in practical applications.

16. (After Box, 1966.) In various chemical reactions, temperature of the reacting solution is one determinant of yield, or efficiency, of a manufacturing operation. To optimize yield, the human operators may be instructed to hold the temperature within narrow limits, taking appropriate action whenever the temperature strays outside these limits. Suppose the question now arises whether the true optimum for temperature lies outside these limits. To answer this question, observational data are collected under normal operating conditions, including temperature and yield, and subjected to multiple regression.

a. By analogy to one-variable regression, why will the temperature–yield correlation be small?

b. Why will there not be a corresponding effect on the b weight for temperature?

c. Granted that the expected value of the b weight for temperature will not be affected from holding temperature within a narrow range, what bad thing will happen to the estimated value of the b weight?

16ANS. a. The temperature–yield correlation will be small because temperature has been controlled within a narrow range.

b. Consider the b weight for temperature as the slope of the yield curve as a function of temperature. This slope will be the same over a small range as over a large range. The cited trouble with the correlation arises because it abandons the physical metric.

c. With a narrow temperature range, the b weight will be less reliable.

17. R and r are both correlations between Y and $\hat{Y}$. For a two-variable regression, why can't R be given a negative or positive sign as with r in Chapter 9?

17ANS. The sign of r is the sign of b_1. With R, b_1 and b_2 might have opposite signs.

18. You have five predictor variables and you wish to test whether X_4 and X_5, taken together, add statsig predictive power, over and above the first three predictors. How do you think this test would be made (extrapolating beyond the information in the text)?

18ANS. Run two regressions, one with the first three predictors, one with all five. The difference in SS_{pred} is the added amount predicted by X_4 and X_5 combined. It has 2 df and its MS may be tested in the usual way.

19. How can you explain the large difference in the predictive correlations for women and men cited in *Clinical Versus Statistical Prediction 1*?

19ANS. Young females are more docile than young males, hence more predictable. (Perhaps "docile" is not entirely appropriate, but you seldom hear a young man called a "goody two-shoes.")

An interesting case arose in early intelligence testing by Binet, namely, why more young males than young females were judged by their teachers as mentally retarded. At that time, psychologists looked to psychophysical indexes, such as reaction time and sensory acuity, as measures of intelligence. Not without difficulty, Binet developed more sensible measures.

20. You have a 3×5 factorial design, but all the data points have been lost except for row 1 and for column 1. The entries in row 1 are 1, 2, 3, 4, 5; the entries in column 1 are 1, 2, 3.

    a. Apply the two-variable regression model of Equation 5b, with the error term omitted to obtain estimates of the lost data. Give the 3×5 table.

    b. On what assumption does the validity of these estimates depend?

20ANS. The two-variable regression model is additive so the lost data can be estimated from the given data. The entry in cell jk may be taken as $j + k - 1$ (or as $(j - a) + (k - b)$, with $a + b = 1$).

The validity of these estimates rests on the additivity assumption—that the interaction residuals are all zero, as is assumed in typical applications of multiple regression (see Equation 5b).

21. To illustrate some difficulties of interpreting regression b weights as measures of importance, consider the formula for b_1 in a two-variable regression:

$$b_1 = \frac{\rho_{Y1} - \rho_{Y2}\rho_{12}}{1 - \rho_{12}^2} \times \frac{\sigma_Y}{\sigma_1}.$$

Here ρ_{Y1} and ρ_{Y2} are the correlations between the criterion Y and the two predictors, X_1 and X_2; ρ_{12} is the correlation between the two predictors; and σ_Y and σ_1 are standard deviations of Y and X_1.

    a. Show that b_1 can be zero even though ρ_{Y1} is not zero. Could this be a serious problem in practice?

    b. Show that b_1 can be nonzero even though ρ_{Y1} is zero. Could this be a serious problem in practice?

    c. Interpret the *Mystery of the Missing Cloud Cover* (Section 16.2.2) in terms of this equation for b_1.

21ANS. a. b_1 will be zero when the numerator of the first term on the right of the given equation is 0. This will occur when $\rho_{Y1} = \rho_{Y2}\rho_{12}$. For example, take $\rho_{Y1} = .36$ and the other two ρs = .6. This could be serious because X_1 would seem to be irrelevant whereas it has a substantial relation to the criterion.

b. If $\rho_{Y1} = 0$, b_1 is proportional to $-\rho_{Y2}\rho_{12}$. Then b_1 will be nonzero even though X_1 is unrelated to the criterion as long as X_1 is correlated with X_2 and X_2 is correlated with the criterion. This could be more serious than case (a) because an irrelevant variable is claimed to be relevant. Multiple regression of diet and health studies, in particular, suffer from this confounding.

c. For the *Mystery of the Missing Cloud Cover*, let X_1 denote fighter opposition, X_2 cloud cover, and Y bombing accuracy. Lacking a measure of cloud cover is equivalent to setting $\rho_{Y2} = 0$ in the above equation. Then b_1 will be proportional to ρ_{Y1}, which was observed to be positive. Had X_2 been measured, b_1 would have been negative.

ANSWERS FOR CHAPTER 17

1. How can a single experiment produce multiple false alarms? Multiple false alarms *and* multiple misses?

1ANS. Three treatment conditions with equal true means allow three false alarms. Extend by adding one treatment condition with different true mean. This added condition may miss being statsig with each of the three first conditions. Four conditions thus allows three false alarms and three misses.

2. In the numerical example of the Student–Newman range test, verify that the final outcome is the separation of the means into the following three subsets, each of which contains means that are not statsig different.

$$\{A, B\}, \quad \{B, C, D\}, \quad and \quad \{C, D, E\}.$$

a. Which two-mean comparisons are statsig?

b. You notice that the means seem to fall into two clusters, $\{A, B\}$ and $\{C, D, E\}$. Use the formula in Note 4.1.1b to get 14.83 ± 7.38 as a 95% confidence interval for the difference between the means of the two clusters. How much confidence do you have that this clustering is real?

2ANS. a. The calculations are an easy application of Equation 1. First, get the denominator, namely, $\sqrt{144/9}$, which equals 4. Each test is then made by dividing the corresponding range by 4 and comparing this q to the q^* given in the text.

Thus, the overall range, $E - A$, yields $q = (20 - 1)/4 = 4.75$, which is greater than the criterial $q^* = 4.04$ for $a = 5$. A and E thus differ.

Next, the ranges $E - B$ and $D - A$ both yield $q = 4.00$, which is larger than the criterial $q^* = 3.79$ for $a = 4$. A and D thus differ, as do B and E.

Both of these subranges yield two comparisons. Only $C - A$ is statsig, with $q^* = 3.50$ for $a = 3$. A and C thus differ.

Two of the subranges are close to statsig, namely, $D - B$ and $C - B$. These differences should not be ignored as they may reflect real trends in the data.

b. This difference between the means of the two apparent clusters may merely be a reflection of the statsig difference between each of A and B and E. What is needed to establish clustering is evidence that the real differences between C, D, and E are small, and similarly for A and B. In view of the width of the confidence interval, these data do not provide this evidence (even had this test not been post hoc).

3. You test four experimental conditions but the overall Anova falls somewhat short of statsig. However, your research assistant points out that your four experimental conditions form a clear a priori rank order and suggests that a linear trend test would be most effective. What do you do?

3ANS. If your assistant is correct that the four conditions do form a clear rank order, you could have decided a priori to assess the linear trend as the probably most powerful and informative analysis. On this basis, you could argue that the linear trend analysis is implicit in the design and didn't need to be stated explicitly beforehand. This argument has considerable merit.

One course of action is to do the linear trend test and, if statsig, follow up with a replication that extends the given study. Another course of action is to write up the results and submit for publication. In this case, it would seem ethical to mention the background for the trend analysis and let the editor and reviewers judge whether your conditions do indeed form a clear, a priori rank order.

4. Show that the one-for-two rule of Section 17.3.2 yields $\alpha_3 = .074$ for $a = 3$ conditions. Compare with Student–Newman procedure.

4ANS. The one-for-two rule of Equation 4 yields $\alpha_3 = 1 - (1 - .05)^{3/2} = .074$ for three conditions, whereas the Student–Newman procedure uses $\alpha_3 = .05$. For the overall range, therefore, the one-for-two rule is somewhat more liberal than Student–Newman. However, both use $\alpha = .05$ at the next step, which tests between adjacent means.

5. In the familywise test of inverted-U shape (Section 17.4.2):

a. What is the null hypothesis? Why is this null hypothesis appropriate?

b. What is the familywise α if H_0 is true?

c. Suppose the expected maximum condition had not been specified beforehand, but selected by inspection of the data. Why exactly would this invalidate the analysis? What modification is needed in this case?

d. Would you have any preference between the Newman–Ryan and the Tukey procedures for this question of inverted U shape?

5ANS. a. The null hypothesis states that all stimulus conditions to the right of the hypothesized maximum condition have equal true means, including the hypothesized maximum condition itself. This is appropriate because the experimental hypothesis asserts that the true means decrease over this terminal limb of the curve.

b. With familywise procedure, the probability of a false alarm is no larger than α.

c. Choosing the maximum condition by inspection of the data capitalizes on chance; this mean will be biased upwards (see *Regression Artifact* in Section 18.4.5), which invalidates the analysis by making the false alarm parameter too large. A valid analysis can still be obtained by excluding this condition and restricting the test to those conditions to the right of the observed maximum condition. Unless prior knowledge about the likely maximum condition is fairly reliable, this test may be preferable. Of course, the choice between these two tests must be explicitly specified beforehand.

d. The Newman–Ryan has more power than the Tukey procedure, so I would prefer it.

6. With one-way design, a statsig range always demonstrates a two-mean difference, whereas a statsig F says only that not all true means are equal with little information about which means differ from which. Hence a range procedure such as Student–Newman might seem preferable to the overall F. Although the range procedures may have a little less power in

most situations, they go beyond the overall F to say which means differ from which.

So: Why not make range procedures standard and forget about overall Anova? As a bonus, much material of previous chapters could be omitted; learning statistics would be much easier. What reason can you see for retaining the overall F for one-way designs?

6ANS. As far as I know, repeated measures designs cannot generally be handled with range tests because of nonsphericity. For independent scores, however, I feel that range tests deserve consideration as the primary tool. With factorial design, range tests could be used for the main effects and F for residuals.

7. You have a 3×4 design for which it is important to make two-mean comparisons for each main effect.

    a. Is it valid to ignore main effect Anovas and instead apply a range procedure to each main effect? What are the pros and cons?

    b. Range procedures are not useful with interaction residuals. Do you think it is valid to use the range procedure for main effects, and then use the overall Anova to test the interaction residuals?

7ANS. a. A range procedure, in contrast to overall F, allows for all possible two-mean comparisons. Since you need two-mean comparisons, Anova seems irrelevant.

b. I believe it is legitimate to use a range procedure for the main effects, and use only the F for interaction residuals. These three tests are statistically independent.

But note: For main effects with three levels, pairwise comparisons of the three pairs of means following a statsig Anova F does not escalate α. With three levels, therefore, Anova is usually superior. The α escalation of the LSD procedure only begins with four levels, as in the example cited in the text.

8. In your undergraduate honors course in experimental analysis, one of your student teams reports a term experiment with four conditions, $n = 11$, in which a test between the largest and smallest means is statsig, $t = 2.37$, compared to $t*(40) = 2.02$. This test seems clearly post hoc to you. In fact, their report explicitly says that the test was made by selecting the largest and smallest means because that was where a real difference was most likely to be!

    a. What is the operative α level of their analysis?

    b. Using $q = \sqrt{2}\, t$, how large would their t have to be to be statsig?

    c. How do you handle this in grading their paper?

8ANS. a. The operative α of their analysis is .20 (see text).

b. From the formula in (b), t would have to be $3.79/\sqrt{2} = 2.68$.

d. Statistical significance is a lesser aspect of teaching at this level; main concern is with the conceptual framework of the experiment, with experimental procedure, and with presentation of data and writeup. After your comments on these main concerns, you could point up pitfalls of post hoc analysis.

9. Justify the assertion ''Obviously, not much credence can be placed in these two results'' (last paragraph of Section 17.4.3).

9ANS. Of the 20 statsig results dredged up, 18 did not hold up in the cross-validation. This underscores the possibility that the two that were statsig in the second analysis were themselves false alarms. If no results were real, one of the 20 would be expected to be statsig in the second analysis. That there were two, not one, is not very convincing.

10. Justify ''When α is reduced for each separate test, the miss rate, β, necessarily increases'' (Section 17.1.3).

10ANS. This is just the α–β tradeoff of Section 2.3.5.

11. P planned six comparisons of which only one was statsig. Why do the five nonstatsig comparisons cast doubt on the one statsig? How should P handle this matter?

11ANS. The failure of five of six planned comparisons suggests that P's knowledge system is not in tune with Nature—that his ''reasonably firm a priori basis'' is eroded a posteriori. If all six comparisons were truly zero, the chance that one would be statsig is about .40. So this one of six is pretty weak evidence. P has two alternatives: Replicate with improved procedure aiming for greater power, especially focused on the one statsig result; shift to some other problem.

12. The critical first step in judging whether Hamilton or Madison had written the disputed papers in *The Federalist* (Section 19.1.3) was to find features of writing style that discriminated between the two in essays of known authorship. Why is there a problem of α escalation here? How would you handle it?

12ANS. Any one essay of Hamilton's and one of Madison's will show chance differences in a fair number of words. Of course, the same may be said of any two essays of either man alone. The search for distinguishing features is a form of data dredging. It picks up chance differences, thereby escalating α.

Do an initial search with one set of essays; reserve a second set for cross-validation.

13. Show that Fisher's (protected) LSD procedure maintains a familywise α for $a = 3$. (Consider all three possible combinations of equal and unequal true means.)

14. With five experimental conditions, each with $n = 9$, Q applies the Student–Newman range test and obtains $q = 3.97$ for the largest mean difference. What will she do now?

14ANS. The q of 3.97 is less than the criterial value of 4.04 cited in the numerical example of the text. Hence the Student–Newman procedure has found no statsig results.

But q is close to statsig, suggestive evidence although not ''beyond reasonable doubt.'' Q will certainly give these data serious consideration in planning her next experiment.

15. In the pitfall with Fisher's LSD procedure cited in Section 17.1.2, why is false alarm parameter for the family nearly 2α?

15ANS. The pit involved two pairs of means, equal within pairs, and greatly different between pairs. Two null hypotheses are true, one for each pair of means, and these two comparisons are independent. This gives two chances of false alarm, so the probability of at least one false alarm is $2\alpha - \alpha^2$.

ANSWERS FOR CHAPTER 18

a–c. Exercises for Sections 18.1–18.3.

a1. Consider *The Mystery of the Missing Cloud Cover* in Section 16.2.2. How does the negative *b* weight for fighter opposition bear on the distinction between process and outcome for assessing importance discussed under *Understanding Regression* b *Weights*?

a1ANS. The negative *b* weight for fighter opposition is relevant to forecasting bomb damage, which is one outcome (among others). It is misleading, clearly, for understanding the processes that determine this outcome.

a2. a. The bigger the real effect, the bigger is the expected *F*, and the smaller is the expected *p*. Thus, *F* and *p* might seem good indexes of effect size. Argue instead that *F* and *p* are poor indexes of effect size because both depend on the number of observations.

a2ANS. In principle, effect size is a property of the phenomenon, unrelated to the number of observations of that phenomenon. A larger *F* may give more confidence that a small effect is real, rather than chance, but a small effect is still a small effect. This reasoning applies equally to *p* (Section 2.4.3).

a3. P and Q do identical experiments, each using different random samples from the same population. P gets *p* = .009; Q gets *p* = .032. Who is better off if:

    a. H_0 is true?   b. H_0 is false?

a3ANS. The real effect is identical for both P and Q, by the statement of the exercise. The difference in their *p* values is only chance. Both are equally well off. Trouble arises only if too much emphasis if placed on observed *p* values.

a4. Multiple regression has been used to assess relative importance of stimulus informers. Mehrabian (1972) showed photographs of women's facial expressions intended to communicate *liking*, *neutrality*, or *disliking*, paired with recorded voices of women saying "maybe" intended to communicate the same three feelings. The three levels of each variable were assigned values of −1, 0, and 1, and used in a two-variable regression analysis. The *b* weights were 1.50 for face, 1.03 for voice. These *b* weights were treated as indexes of importance; facial expression was thus considered 50% more important than voice tone.

    a. What is the argument to treat the *b* weights as measures of importance?

    b. What is the fatal flaw in the argument of (a)?

    c. With a 3 × 3 design, could Anova measure relative importance?

a4ANS. a,b. The *b* weight for each separate variable is the slope of the linear regression line. If *b* = 0, the variable has zero importance for the response. Larger *b* means greater impor-

tance. Unlike *r*, moreover, *b* is nearly independent of the range of *X*.

The fatal flaw is the implicit assumption that − 1, 0, + 1 are comparable between the two variables, that is, that these numerical scales have the same unit for face and voice. Equal units is unlikely because the actual informers were chosen arbitrarily.

c. Factorial design would allow use of the relative range index of Equation 4, which does not require stimulus values. However, importance is still confounded with arbitrary choice of stimuli.

The question of assessing relative importance has arisen in diverse fields; people expect it to have a definite answer, but often it does not. A number of writers have used multiple regression coefficients, as in the face–voice comparison by Mehrabian (1972) used in this exercise, and others have proposed similar indexes, as with analysis of emotions based on facial expressions (see Anderson, 1989, p. 166). Perhaps without exception, these attempts to compare importance are invalid. Valid comparisons are sometimes available with functional measurement (Chapter 21).

a5. Suppose the additive model holds so that each cell mean can be written as $\alpha_j + \beta_k$ (ignoring error and the overall mean). Show that the row means may be written as $\alpha_j + \bar{\beta}$. Show that the sum of the ranges for the row and column means equals the range of the cell means. Relate this to the relative range index of Equation 4.

a6. a. From Equation 1, show that $|\mu_1 − \mu_2| = 2\sigma_A$ for two groups.

    b. From (a) and Equations 2–3, show that $d = 2f$.

a6ANS. a. For two groups, $\bar{\mu}$ lies halfway between μ_1 and μ_2, so $|\mu_1 − \bar{\mu}| = |\mu_2 − \bar{\mu}| = \frac{1}{2}|\mu_1 − \mu_2|$. Equation 1 becomes $\sigma_A^2 = (1/4)|\mu_1 − \mu_2|^2$; so $2\sigma_A = |\mu_1 − \mu_2|$.

a7. P and Q have been independently funded by an international pharmaceutical corporation to develop a chemical intended to decrease family quarreling and increase family happiness. Both report success, but the development costs, especially getting FDA approval, are so huge that only one can be developed. To decide which one, the corporation secures an independent evaluation of each chemical, compared with a placebo control, $n = 31$ families, with equivalent response measures for each chemical. The results show 99% confidence intervals of 8 ± 8 for P, 4 ± 4 for Q.

    a. What are the *p* values for P and Q? What do they imply/not imply?

    b. Does it make any difference which chemical the corporation decides to develop? If not, why not? If so, which one?

a7ANS. a. For both P and Q, *p* = .01. These *p* values imply the observed effect is unlikely—if the null hypothesis is indeed true. They do not imply the effect is equally important for each chemical. Nor do they imply the effect is equally large.

b. P's mean of 8 is larger than Q's mean of 4. If a choice had to be made on these data, it would be P. (The exercise specified "equivalent response measures.")

b1. Why *should* the c_j in a contrast sum to zero? To answer this question, assume the null hypothesis is true and see what will happen if the c_j do not sum to zero. First consider the case of $a = 2$ conditions.

b1ANS. b. Suppose that all true means equal some nonzero constant, k. Any contrast, $\sum c_j \mu_j$, then equals $k \sum c_j \neq 0$. This absolutely must not be.

b2. Use the comparison coefficients of Table 18.1 to show that the interaction residuals are identical in all three panels of Figure 5.4.

b2ANS. In contrast form, the interactions in the three panels of Figure 5.4 are:
$$12 - 15 - 9 + 18 = 6, \quad 12 - 15 - 0 + 9 = 6, \quad 12 - 15 - 5 + 14 = 6,$$
from left to right.

b3. You wish to test the mean of two groups against the mean of three other groups, all with equal n. Give the normalized c_j and write the expression for the confidence interval.

b3ANS. Normalized contrast coefficients are 1/2 for the two groups and $-1/3$ for the three groups.
$$\left[(\bar{Y}_1 + \bar{Y}_2)/2 - (\bar{Y}_3 + \bar{Y}_4 + \bar{Y}_5)/3 \right] \pm \sqrt{5/6} \times t^* \times \sqrt{MS_{error}/n}.$$

b4. Given $Y = 1, 3, 5, 5$, for $X = 1, 2, 3, 4$, with $n = 6$ and $MS_{error} = 8$.

    a. Show that the power for linear trend of this curve is .75.

    b. Show that the power for the overall F is .57.

    c. Plot the curve. What is the moral of this exercise?

b4ANS. a. For the linear trend, using $\{-3, -1, 1, 3\}$ as the c_j:
$$\sigma_{error} = \sqrt{\sum c_j^2} \sqrt{MS_{error}} = \sqrt{20} \sqrt{8} = 12.649.$$
$$\psi = -3 \times 1 - 1 \times 3 + 1 \times 5 + 3 \times 5 = 14. \quad \sigma_\psi = \sqrt{\psi^2/2} = \sqrt{98} = 9.899.$$
$$f = \sigma_\psi / \sigma_{error} = .783. \quad \phi = .783 \sqrt{6} = 1.92. \quad \text{Power} = .75.$$

b. $\alpha_j = -2.5, -.5, 1.5, 1.5$, for the successive values of X.
$$\sigma_A = \sqrt{\sum \alpha_j^2/4} = 1.658.$$
$$f = \sigma_A / \sigma_\varepsilon = 1.658/\sqrt{8} = .586. \quad \phi = .586 \sqrt{6} = 1.435.$$
Power = .57.

b5. From Equation 12, show that multiplying all c_j by a positive constant k increases the width of the confidence interval by a factor of k. Is it good or bad for the confidence interval to change width like this?

b6. Justify the statement of the text that normalization of contrasts with a 2^p design requires a divisor of $1/2^{p-1}$.

b6ANS. For a 2^p design, 2^{p-1} of the c_j are $+1$ and 2^{p-1} are -1. QED.

b7. You have a quantitative model which predicts that $\mu_1 - \mu_2$ equals some nonzero constant, c. How do you test this hypothesis?

b7ANS. You can test H_0: $\mu_1 - \mu_2 - c = 0$ with a confidence interval, $\bar{Y}_1 - \bar{Y}_2 - c \pm \sqrt{2} \, t^* \sqrt{MS_{error}/n}$.

c1. In *Rank and Rating Metrics*, what is the advantage of rating over ranking?

c2. Consider the cubic curve, $Y = X^3$. Suppose $X = -1, 0, +1$. How much of the SS lies in the linear component? (Plot a graph; no calculation is needed or wanted.) What is the purpose of this exercise?

c2ANS. For $X = -1, 0, +1$, $X^3 = X$. The graph is thus a perfect straight line, so all the SS lies in the linear component.

    This exercise shows that the same mathematical function can exhibit different shapes depending on the values of X. Hence the shape of observed data, even errorless data, is not a dependable clue to the shape of any underlying function—another manifestation of the process–outcome distinction.

c3. Consider the pure quadratic curve, $Y = X^2$, with $X = 0, 1$, and 2. Show that the total SS is $8.667n$, of which 92% lies in the linear component, only 8% in the quadratic component (the quadratic c_j are 1, -2, 1). What is the moral?

c3ANS. Curve shape is an ambiguous concept for it depends on the range of X.

c4. Subjects receive four trials per day in an emotional adaptation task, using one emotional arousal condition each day. Each subject serves under three different emotional arousal conditions on three successive days. You wish to study adaptation of arousal, as it occurs within and between arousal conditions.

    a. Suppose you plan to use the overall repeated measures Anova (Chapter 6). Write out Source–df and say which sources test your hypotheses.

    b. In looking at your answer to (a), you wonder whether you should pass by the overall Anova and instead test linear trend. What do you think?

    c. Before you decide to plan any analysis around the linear trend of (b), you feel you had better be sure you can actually make this calculation. Show how to use the contrast analysis of Section 18.2.1. It may help to begin with a numerical example.

    d. What other variable needs to be controlled in this design?

c4ANS. a. This is an $S \times A \times B$ design, with $A = $ trials and $B = $ arousal conditions. Your adaptation hypothesis implies a trend of decreasing emotional response across successive trials. The A effect tests for changes across trials within days, averaged over arousal conditions. The B effect tests for difference between arousal conditions, averaged over trials. The AB term tests whether the trials effect depends on arousal condition. Your two hypotheses are thus tested by A and AB.

b. Adaptation is likely to lie mainly in the linear component, so the linear trend is an attractive score. It can localize the expected effect, which the overall F does not, and it is likely to be more powerful.

c. A linear trend would be calculated for each subject for each condition, by multiplying each subject's scores on the four successive trials by -3, -1, 1 and 3 (or $-3/4$, $-1/4$, $1/4$, $3/4$, with normalization). This reduces the four scores on each day to a single score, ψ_i.

Each subject thus has three trend scores, one for each arousal condition, which would be analyzed in the Subjects $\times$ Conditions design. A statsig F_{mean} would indicate real changes across trials, averaged over conditions. A statsig $F_{conditions}$ would indicate that the within day trials effect differs across conditions.

d. The rate of adaptation may well change over days. The design as given confounds conditions with days, which seems undesirable. Conditions could be balanced over days with a 3×3 Latin square.

c5. Given $Y = 1, 5, 7, 8$ for $X = 1, 2, 3, 4$, with n observations in each condition. Graph this curve, fit a straight line by eye, and guess what proportion of SS_A will be in the linear component. Check your guess by calculating the two SSs. Comment.

c5ANS. $SS_A = 28.75n$, $SS_{linear} = 26.45n$, so 92% of the SS lies in the linear component. A pretty nonlinear curve can still have most of its variance in the linear component. Further, linear trend may well be more powerful than overall Anova.

c6. Find a published curve in some area of interest to you. Suppose the response measure is only a monotone scale. Graph two other curves of quite different shape that would result from monotone transformation of these data and discuss whether the interpretation would differ in the three cases. How much reason do you have to think that the published curve is more valid than your two alternatives?

d. Exercises on Regression Artifact

d1. College students tend to have children less intelligent than themselves.

   a. Why is this guaranteed by statistics?

   b. What are its implications for your own parenting?

d1ANS. a. The parent–child correlation in intelligence is substantially less than 1. Since college students are above average in intelligence, the regression effect guarantees that their children will on average be less intelligent.

b. Your children are likely to be less intelligent than you or your spouse. They may even be below average. If you expect too much of them, you may misshape their lives.

This question stems from a large lower division course on personality that I taught, in which I assigned a term paper on "How I would raise an average male child." Some of the female students had strong emotional objections to this assignment, saying that *they* certainly would not have an average child.

(The great breadth of responses in these term papers was an eye-opener to me. One male said he would pad his child's room with mattresses to keep him from hurting himself. Quite a number of females expanded on the idea that they would immerse their child in lavish total love. I addressed a special lecture with the title "Love is not enough," emphasizing the need for application of psychological knowledge about parenting. Since then, I have come to better realization of how little is known about parenting and how extremely neglected is this most important research area in psychological science.)

d2. In Chapter 9, Galton's equation showed that sons of taller/shorter fathers tended to be shorter/taller—closer to the population mean in both cases. Some writers have argued that the population as a whole is moving toward the mean, with increasing homogeneity over successive generations.

   a. What will happen if we compare the heights of taller (shorter) sons with the heights of their fathers?

   b. What does your answer to (a) suggest about systematic regression to the population mean over successive generations of humans?

   c. What does the hypothesis of systematic regression imply about the differences between the distributions of height across successive generations? What statistic is appropriate for empirical tests?

d2ANS. a. Galton's equation should still hold; just interchange father and son in the equation. Thus, taller sons will tend to have fathers shorter than themselves; and similarly shorter sons will tend to have fathers taller than themselves.

b. The argument in (a) shows the same pattern of data in the backward direction as in the forward. To argue for increased homogeneity from the forward pattern implies the same for the backward pattern; that would mean the population must already have become homogeneous, which is manifestly not true.

c. In contrast to most statistical analyses, the mean is not relevant. What is relevant is the variance of the population. If the homogeneity argument is correct, as is logically possible, then the population variance will decrease over time.

(This example is covered in Example 261A, p. 261, of Wallis and Roberts (1956), who cite other examples, such as claims for the triumph of mediocrity in business, as well as a speech by Woodrow Wilson, who commented on the "rise out of the ranks of unknown men" without falling into the black pit of the regression artifact.)

d3. (After Wallis & Roberts, 1956, pp. 258-261.) Many sports fans are avid connoisseurs of sports statistics. Bill McGill suspects major league batters are becoming mediocre, with the good ones becoming poorer, the poor ones better. As one illustration from the 1954 baseball statistics, Mays (Giants) led the major league with a batting average of .345, whereas Williams (Giants) came in last at .222. Sure enough, one year later, Mays dropped to .319 and Williams rose to .251.

Bob Boynton, on the other hand, suspects the good get better and the poor get poorer. He looked at the same two years as McGill. To illustrate, Kaline (Tigers) led the major league in 1955 at .340, whereas O'Connell (Braves) trailed at .225. Sure enough, one year earlier, in 1954, Kaline and O'Connell averaged .276 and .279, respectively.

a. Give a psychological rationale for McGill's hypothesis.

b. Give a psychological rationale for Boynton's hypothesis.

c. Boynton and McGill recognize that their results look contradictory. Of course, they realize anything can happen with selected examples. To know what is really going on, they must look at mean scores for all above average players and for all below average players. They do this. What do they find? And why?

d3ANS. a. One psychological rationale for McGill's hypothesis is that players with lower batting averages are more motivated and try harder. Some players with higher batting averages, on the other hand, are inclined to take it easy.

b. One psychological rationale for Boynton's hypothesis is that good players have higher upper limits and improve more with practice. Further, their success gives them higher morale. Poorer players, in contrast, are just naturally poorer and have less room for improvement. Further, they tend to suffer from lower morale and high stress. (A nonpsychological reason may be added, namely, that they get fewer opportunities to play and improve their game.)

c. By looking at the mean scores for all players, McGill and Boynton will find the same trend they found in the selected examples. The regression artifact operates in both cases. But its direction depends on whether the classification into better and poorer is made in the first or the second of the two years.

d4. In the second paragraph of Section 18.4.5, give the reasoning of "Similar reasoning applies to subjects with equal true scores near the boundary."

d5. An alternative way to evaluate Head Start would be to match each Head Start child with a comparison child of same age, gender, and socioeconomic status. Both groups would be tested on vocabulary, say, or IQ. The great advantage is that this matching can be done after the Head Start program has been completed, thereby allowing a test in later years for permanence of the Head Start effect. How can the regression artifact undercut this comparison?

d5ANS. Presumably the Head Start children will be matched with nondisadvantaged children who are superior on the various measures. Nondisadvantaged children whose scores on the test used for matching are less than their true scores will tend to be selected as matches. On the critical test, they will regress to their higher scores, thereby making Head Start look harmful. This is one of six such artifacts discussed by Campbell and Boruch (1975) in connection with the Head Start program.

d6. In the example of comparing slow and fast learners in Section 18.4.5, justify the statement "But because of unreliability, the observed score on the criterial trial is higher than the true score." How could this be tested empirically?

d6ANS. To show empirically that the observed score on the cri-

terial trial is higher than the true score, add one or a few trials past the criterial trial. To the extent that unreliability is present, the score on the criterial trial will stick out above the curve through the other trials. If the learning curves for different subjects are aligned at the criterial trial for one perfect list performance, the next trial will show a marked downturn.

An "endspurt" is a less obvious consequence of aligning the individual learning curves at the criterial trial. The increment from the previous trial to the criterial trial will be larger than increments on trials just preceding. Some people thought to interpret this endspurt to mean that subjects sensed they were near the goal and tried harder. The artifactual nature of this endspurt was clarified by A. W. Melton: "The end-spurt in memorization curves as an artifact of the averaging of individual curves." *Psychological Monographs*, *47*, 1936, Whole no. 212, 119-134.

d7. (After Rulon, 1941.) Whether progressive education does better than conventional instruction has been of concern to many investigators. In the absence of randomized experiments, some investigators have sought to compare different groups by matching subjects. One such study compared two groups: One group had conventional education through the first four grades; the other group spent the third and fourth grades in a program of progressive education. The assiduous investigator matched the two groups of children on six variables: mental age, chronological age, physical ability, social qualities, home surroundings, and educational achievement. The question was how these two groups compared at the end of the fifth grade. "In every [academic] subject it was found that the group with the background of progressive education did better than the group with the conventional background."

a. How could the regression artifact enter here?

b. How does this example differ from the Head Start comparisons cited in the text and other exercises?

c. The question of effective education has highest importance. What way can you suggest to avoid the ambiguity of the matching method?

d7ANS. a. It seems likely that those children who received the progressive education had better home backgrounds than those children who received conventional education. The fact that their parents had enrolled them in the progressive program suggests the parents had higher education, higher income, more books, and so forth.

Consider a matched pair of children. The scores on which they are matched will be in error through unreliability of response or other measurement. Granted that the one group has higher true scores than the other, matching will tend to select those children in the progressive group whose observed scores are lower than their true scores. Similarly, matching will tend to select children in the conventional group whose observed scores are higher than their true scores. In the tests at the end of the fifth year, regression will reveal this difference in true scores.

b. This differs from the Head Start and other typical comparisons in that the regression effect operates in favor of the superior group.

c. Multiple regression to "control" uncontrolled variables is a step above matching, but typically suffers similar fatal con-foundings (Section 16.2). Quasi-experimental design suffers in the same way (Section 15.5). Causal models can be effective, but they depend critically on prior knowledge about causality that is unlikely to be available in education or in most social situations and at best suffer from partial measure bias.

When randomized experiments are at all feasible, alterna-tives are usually pointless. A model comes from randomized experiments in medicine, which are not uncommon, and which often use thousands of subjects—large Ns that are needed for many outcome experiments on education. If patient consent can be obtained in medical experiments, parental consent should often be obtainable for experiments on education, in which the worst treatment is likely to be the status quo. And if money can be obtained for such medical experiments, money should surely be obtainable for reseach on education.

The dismal state of education science is a black mark on the university system (see Notes 23.2a,b, page 781).

e. Exercises on Section 18.4.

e1. The editor of *Psychophysiology* asks you to review a submit-ted article on emotion, in which the investigators measured heart rate, galvanic skin response, and finger blood volume for each subject. Two groups of subjects were run to test a theoretical prediction from functional theory of emotion. You consider their experimental procedure to be sound and their theoretical logic seems reasonable. The Anovas show $p < .20$ for each of the three measures. "Although these separate tests are not stat-sig," say the investigators, "they are all in the predicted direc-tion. The chance that all three are simultaneously less than .20 is $.20^3$, which equals .008. This seems definite evidence for a real effect."

The editor has asked you for a review that can be forwarded to the authors. Do you recommend acceptance, revision, or rejection?

e1ANS. The probability argument is erroneous. It assumes the three tests are independent, whereas they are correlated measures on the same subjects. Manova, however, will combine the three measures in a single analysis that extracts the independent infor-mation. The authors should apply Manova, which might salvage their experiment.

e2. The test whether vitamin–mineral treatment would raise nonverbal IQ cited in Section 17.4.2 was meritorious in using several measures of nonverbal IQ. But the data analysis was shameful, for it tested each measure separately.

   a. Why was it meritorious to use several measures of nonverbal IQ?

   b. Why was the data analysis shameful?

   c. How could Manova help?

e2ANS. a. Use of several measures was meritorious in two ways. Most important, it allows nonverbal IQ to have more than one factor, much as with standard verbal IQ. Even if there is only one factor, moreover, the best measure of it would probably be unknown. Each of the multiple measures may thus contribute

to an overall, composite measure that may have more sensitivity to the experimental manipulation.

b. The data analysis failed to allow for α escalation from multi-ple tests (87 tests all told). This is shameful because this danger is well known.

c. Manova can help by integrating the multiple measures into one, so only one false alarm is possible. (This somewhat simplifies the actual study, as some of the 87 tests were made between subgroups, not for 87 separate measures.)

e3. For the data of Table 18.2a, verify the MS_{error} cited in the text and use it to show that $F(3, 3) = 8.72$. Do a log transform and apply Anova. Discuss implications of these two analyses.

e3ANS. The error term is the SA interaction residual, which may be computed in the usual way. (Note that MS_{SA} equals the sum of squared interaction residuals listed in Table 18.2b, divided by the 3 df; 3 df remain because the residuals sum to 0 in each row and in each column.

Anova of the logs yields $F = 100$, comfortably greater than $F^*(3, 3) = 9.28$. These data are purely multiplicative if .2 is subtracted from each value. Anova on $\log(Y - .2)$ yields an F close to 140. It may thus be useful to allow for a zero point in similar way when planning a log transform.

This artificial example may be overly suggestive. In prac-tice, only modest benefits should be expected from transforma-tion. How much benefit can be expected in particular empirical situations can be assessed by doing the analysis in both ways.

e4. Justify the assertion of the text in Section 18.4.6 under *Nature of Error Variability* that "We gain confidence in a real effect to the extent that the curves for all the subjects show a common pattern."

e4ANS. The curve for each subject portrays the relative effects of the treatments on that subject. Similar real effects for dif-ferent subjects will appear as similar curves; zero real effects will yield only chance similarity of curves for different subjects. Hence similar curves argue for similar real effects. (It is possi-ble, of course, to get nonsimilar curves from real effects that differ across subjects.)

e5. You have a 3×5, single subject design with two replica-tions, using treatment randomization to obtain independence. You plan to give preliminary practice and you are confident that any residual practice effect will be negligible. However, you anticipate that other investigators in your field will be concerned about practice effects. Accordingly, you decide to present the first and second replications in succession, using treatment ran-domization separately within each replication. Differences between the two replications will then provide a measure of practice effects.

   a. Write out Source–df for Anova, treating replications as a third factor with two levels. What is disconcerting about this Anova?

   b. An alternative design would randomize all 30 trials en bloc, without regard to successive replications, which there-fore would disappear as a factor in the design. What advan-tage does this have? What disadvantage?

c. How could you apply partial analysis?

e5ANS. a. This is a standard three-factor Anova: 2 and 4 df for your two experimental variables; 1 df for replications; and corresponding df for the four interaction residuals. What is disconcerting is that there are no df for error because $n = 1$.

b. The advantage of en bloc randomization is that it eliminates replications as a factor and thus provides 15 df for error. The disadvantage is that you obtain no information about possible effects of practice.

d. With partial analysis, the main effect of replications could be used to assess practice effects. That would leave 14 df for error, obtainable by summing the SS and df for all interaction residuals involving replication. A large effect of replications, being contrary to your expectations, would warn that the experiment itself may need replication.

e6. Some reports have followed a statsig Manova with separate Anovas on the separate measures. To the objection that the separate Anovas allow false alarm escalation (Section 4.2.2), they point out that Manova maintains the assigned α, and claim that this justifies the follow-up tests. This claim is examined here.

Suppose that four measures were used, one with a huge difference between groups, the other three with zero difference between groups. For simplicity, assume all four measures are uncorrelated and independent.

a. What is the probability of a statsig Manova?

b. What is the probability of a false alarm on the separate Anova of each useless measure?

c. What is the probability of a false alarm on at least one of the Anovas on the three useless measures?

d. How will the above answers change if all three useless measures are perfectly correlated?

e6ANS. a. The probability of a statsig Manova is virtually 1 because of the huge difference on one measure.

b. The probability of false alarm on the separate Anova of each useless measure is α.

c. Given independence, the probability of a false alarm on at least one of the three useless measures equals 1 – the probability of no false alarm on any one: $1 - .95^3 = .143$.

d. The answers to (a) and (b) do not change. The answer to (c) changes to .05; all three useless measures will be statsig or not at the same time. It may look impressive to see all three statsig as though that surely could not happen by chance, but this impression rests on an implicit assumption that the three are independent. Something similar will happen if they are partly correlated.

e7. In the fourth paragraph of Section 18.4.4, justify the statement about a clear danger of a too-small error term. Why is pooling dangerous in this situation?

ANSWERS FOR CHAPTER 20

a. Exercises on Addition Model

a1. How can you test goodness of fit for Piaget's centration hypothesis for the time judgments of Equation 1c of Section 20.1 (see Exercise 5.2):

    a. With Anova?

    b. With standard regression analysis of Chapter 9 or 16?

    c. Which seems preferable?

a1ANS. Piaget's centration hypothesis implies that only one of speed and distance has a real effect. The regression trend on 1 df should be more powerful that Anova main effect on $a - 1$ df.

a2. In Figure 20.3 on the size–weight illusion:

    a. Why would it be gauche to report the F for gram weight?

    b. Would it be gauche to report the F for size?

    c. Suppose the main effect for size was not statsig. How would this bear on the purposes of the experiment?

a2ANS. a. No experiment is needed to show that gram weight has a real effect on judged heaviness. To report the F would betray total naivety.

b. That size has a real effect is clear from visual inspection of the figure. Its magnitude is shown by the slope of the curves, and its reliability is indicated by the regularity of the pattern of curves. However, a within subject, two-mean confidence interval to measure reliability would seem desirable.

c. Deviations from parallelism would be just a fraction of the main effects. If the main effect of size was not statsig, therefore, adequate power to detect nonparallelism would be lacking. Indeed, a smallish effect of size, even though statsig, would probably not provide adequate power.

Note that the one-point deviation from parallelism in Figure 20.3 was enough to produce a statsig interaction residual. This demonstrates adequate power.

(This experiment, it may be added, probably erred in using cylinder height to vary size. Height varies only in one dimension. A previous experiment had used cubes as stimuli, which vary in three dimensions, and had obtained a substantially larger illusion. The size effect in Figure 20.3 was not as strong as it could have been.)

a3. Restate in your own words the reasoning of the last paragraph of *Cognitive Development* in Section 20.1.3, beginning "The conceptual implications of this result are more important than the algebraic rule itself."

a3ANS. The speed variable is conceptual, not simply perceptual, because it is defined symbolically by a picture of an animal. Yet this picture has a metric function comparable to the perceptual metric cue of distance. Finding an additive integration suggests that these concepts are real to the children. Wilkening's experiment thus revealed notable conceptual capabilities of young children, previously denied in Piagetian theory.

No less striking are other capabilities required in this task. In particular, the child has to construct a complex assemblage within which the judgments are made (Note 20.1.3b).

Mathematical models are important for their conceptual structure, in this case, for understanding development of knowledge of the external world. They may also be useful for studying the fundamental problem of assemblage.

a4. For the 3×3, Distance $\times$ Speed experiment considered in Section 20.1:

a. Make up hypothetical data for a single subject that follow the postulated subtraction model exactly. Add an error of $\pm c$ in each cell, where c is a constant equal to about 10% of the range of data, choosing the $\pm$ sign at random for each cell. Make two replications. Apply Anova and comment.

b. Make similar data for the physically correct division model and compare the factorial graphs for both sets of data by graphical inspection.

c. Test the subtraction model with the data generated by the division model using visual inspection and also Anova. Comment.

d. The cell means predicted by the subtraction model are equal to (row mean + column mean − overall mean). Apply this subtraction model to the division data of (b). Graph these predicted values as a function of the observed values and compute the correlation. Discuss in relation to Figure 20.5 and your analysis in (c).

a4ANS. c. Anova should show a statsig interaction residual for the nonadditive division model, at least if both factors cover a substantial range. If the three levels of speed were close together, of course, deviations from parallelism would also be small.

d. This exercise illustrates weak inference. A very high correlation will be found between the data generated by the division model and the cell means predicted by the subtraction model. But the Anova interaction of (c) should be statsig. As in the weak inference example of Figure 20.5, the correlation fails to look at what is important for the purpose at hand, namely, the deviations from prediction.

a5. Serial belief integration can be studied using serial-factor design, in which each serial position constitutes a factor in the design. The main effect of each factor then measures the effect of the informer stimulus at the corresponding position. To illustrate the idea, consider three serial positions, each of which may present an informer of value 0 or 100, that is, against or for some belief issue. Subjects judge their belief only once, after all three informers have been presented, based on all the given information. Asssume that the response to sequence i is the weighted average, $\sum_{i=0}^{i=3} \omega_i \psi_i \div \sum_{i=0}^{i=3} \omega_i$, where ψ_i is the value of the informer at position i and ω_i is its weight. Subscript 0 indicates the initial belief, which can here be ignored (see further Exercise 21.c2).

There are eight possible sequences of informers, denoted by sequences of 0s and 1s. You get the following mean responses to the eight sequences:

1 1 1	90.91	0 1 1	72.73
1 1 0	54.55	0 1 0	36.36
1 0 1	63.64	0 0 1	45.45
1 0 0	27.27	0 0 0	9.09

a. Plot these data to verify that they satisfy the parallelism property.

b. If the averaging model holds, then the weight at each serial position is proportional to the main effect of that serial position. On this basis, show that the weights for the three successive positions stand in the ratio 2:3:4.

a5ANS. a. All three two-factor graphs show parallelism. In addition, a three-factor graph should be plotted to show that both two-factor graphs are geometrically congruent.

b. The main effects for the three successive serial positions are 18.2, 27.3, and 36.4. These stand in the ratio 2:3:4. Because the informer values are equal across serial positions, the weights are proportional to these main effects.

a6. Deservingness is a prominent social concept, especially in judgments about rewards and punishments. Two determinants of judged deservingness are what a person achieves and what the person needs. Design an experiment with children to test whether these two determinants are integrated by an adding-type rule, including reasonable amount of procedural detail. For example, children might be asked to role play Santa Claus in distributing toys at Christmas time. Make up hypothetical data and test the given model.

a6ANS. The Santa Claus scenario is intended to set the task in a framework familiar to many children, although it is not universal. Various experimental precautions are in order: putting the children at ease at the beginning; using a graphic rating scale, such as a rod with 20 rings to represent the maximum of 20 toys for the most deserving children; representing need by some realistic manipulation, such as the number of toys the story child already has; representing achievement by amount of some good deed the child has done, as with helping mother with housework; both factors perhaps preferably in visual form. At least three levels of each factor seem desirable; a 2 × 2 design has too little diagnostic power, especially power to diagnose causes of deviations from the model. The study by Anderson and Butzin (1978) used similar procedures, and gave provisional evidence for addition. I am not familiar with any further work.

Perhaps more interesting would be data that deviated from the model in some likely manner.

a7. Some writers attempt to assess goodness of a mathematical model in terms of the proportion of variance it accounts for. Most frequent is $r^2 = SS_{pred}/SS_Y$. Equivalently, $r = \sqrt{r^2}$ may be used, that is, the correlation between the observed values and those predicted by the model.

An alternative is to use proportion of variance not accounted for, such as $1 - r^2$ or RMSD $= \sqrt{1 - r^2}$. If RMSD is less than .05 (or 5%), say, the model fit is deemed good.

a. For prediction and other outcome analyses, which approach seems preferable?

b. For analysis of integration process, which approach seems preferable?

c. For process analysis, why are both approaches questionable in light of Figures 20.4 and 20.5?

a7ANS. a. For outcome analysis, the first approach seems preferable because it looks at the amount of predictability. In the multiple regression equation for predicting success in graduate school, for example, there is no question that the prediction is far from perfect. But the usefulness of the equation depends on its predictive power, not its imperfections. In an abstract sense, r^2 and $1 - r^2$ are equivalent, but not in a practical sense.

b. For process analysis, it is usually preferable to look at the deviations from prediction. These are what shed most light on process analysis. As the examples of weak inference show, high predictive power can hide serious discrepancies in the model.

c. For process analysis, both approaches are questionable because they do not test the essential question: Are the *deviations* from prediction real?

a8. The logic of placebo control rests implicitly on a rudimentary mathematical model, a form of theory control (Section 8.2.5).

a. Write down this model, defining your terms.

b. What empirical assumption is critical to this control?

c. How robust is this model against nonadditivity?

Note to Instructor. Despite its conceptual simplicity, this problem asks for a way of thinking that is unfamiliar to many first-year students. It may be helpful to provide some cues, as through preliminary class discussion.

a8ANS. a. Let P represent the effect of the placebo treatment (including suggestion, temporal changes, and so forth); let E represent the effect of the experimental treatment; and let μ represent the base level of response in absence of any treatment. The model involves two equations:

$$Y_P = \mu + P;$$
$$Y_E = \mu + P + E.$$

Subtraction yields $Y_E - Y_P = E$, the standard measure of the experimental effect.

b. The critical assumption of this model is that P is the same in both conditions. This assumption is sometimes hard to fulfill experimentally, as when double blinding is not possible.

c. The model is robust against deviations from additivity, for example, if the experimental treatment amplifies, or multiplies, the placebo effect. With just two conditions, any effect can arbitrarily be represented as an additive effect.

a9. In Figure 20.1, it might be argued that the parallelism for the 5-year-olds is an artifact—that half the 5-year-olds center on distance, half on speed.

a. Show by numerical example that this form of the centration hypothesis implies parallelism in the factorial graph of the mean data, averaged over children (disregarding response variability).

b. How can you analyze the data to test this artifact interpretation?

c. Give algebraic proof for (a), allowing an arbitrary proportion π of children to center on Distance, $1 - \pi$ on Speed.

a9ANS. b. To test for this averaging artifact, do single subject Anovas. By hypothesis, each child centers on one dimension, and so should have only one statsig main effect.

a10. For the regression model of Section 20.1.1:

a. How many df are there for main effects of distance and speed?

b. Where do the 9 df for error come from?

c. How many df for deviations?

d. What are the two components of the deviations term?

a10ANS. d. The deviations term includes the nonlinear components of distance and speed as well as the Anova interaction. Only the latter, however, is a proper test of the model. In particular, nonlinear components of distance and speed refer to the physical metrics for these variables, which may be nonlinearly related to the psychological metrics. These nonlinear effects have no bearing on the model. Of course, the regression analysis can be extended to test only the interaction residual, just as in Anova.

b. Exercises on Signal Detection Model

b1. Discuss differences and similarities between Figures 20.6 for decision theory and Figure 2.1 for significance testing.

b1ANS. The two probability distributions in Figure 20.6 are direct analogues of those in Figure 2.1. H_0 true in Figure 2.1 corresponds to Noise in Figure 20.6. The X_N distribution in Figure 20.6 is thus the direct analogue of the H_0 true distribution in Figure 2.1.

Signal present in Figure 20.6 corresponds to real effect in Figure 2.1, that is, to H_0 false. The X_{SN} distribution in Figure 20.6 is thus the direct analogue of the distribution for H_0 false in Figure 2.1. The two figures are thus conceptually equivalent.

Signal detection experiments, however, differ from most other experiments in two respects. First, the signal is controlled, so which distribution is operative on each trial is known. Second, numerous observations are usually available from *both* distributions. In a significance test, in contrast, only one observation is typically available (e.g., a single F ratio), and which distribution it comes from is unknown.

b2. Section 20.3.3 states that "the function of the threshold is taken over by the decision criterion."

a. What is the "function of the threshold?"

b. How does the concept of criterion differ from the concept of threshold?

c. What experimental manipulation cited under *Goodness of Fit* in Section 20.3.2 implies that a concept of criterion is necessary?

b2ANS. a. The concept of threshold arose to explain why some signals did not evoke a response. Thus, a person with more acute hearing was thought to have a lower threshold, able to detect auditory events unnoticed by other persons. With the threshold concept, the stimulus had to exceed some "threshold" value to be detected; that was its function.

b. The concept of criterion differs from the concept of threshold in the representation of the stimulus. The threshold concept implies an all-or-none representation, in which the effective stimulus is either detectable or not. The criterion concept allows a continuous representation, in which the effective stimulus varies in magnitude on a continuous scale.

c. Subjects can be instructed to adopt a laxer criterion, so a concept of criterion seems necessary, even with threshold theory.

b3. To study memory under hypnosis, subjects were shown a videotape of an auto crash in their normal state. They were then tested with leading questions such as "Did you see the stop sign?" in both normal state and under hypnosis. These are called "leading" questions because the wording implicitly suggests that the correct answer is "yes." Actually, all questions were false. Subjects gave more "yes" answers under hypnosis than in their normal state. (For a review of this issue, see Klatzky & Erdelyi, 1985.)

a. Can we conclude that memory retrieval is poorer under hypnosis?

b. What fatal flaw disallows application of SDT to this experiment?

c. Suppose hypnosis has negligible effects on memory retrieval. How might this negative answer be important for the legal system?

b3ANS. a. Since all questions were false, all the "yes" responses were false alarms. Hence we can conclude that hypnosis causes subjects to guess "yes" with higher probability under threshold theory, or equivalently, to use a laxer criterion under SDT. But this is all we can conclude, and this says nothing about memory retrieval.

b. The fatal flaw is the lack of any questions that have "yes" as correct answers. We need two kinds of trials, corresponding to the Noise and Signal-plus-Noise distributions of Figure 20.6. This experiment used only Noise trials.

c. Persons who have observed some crime are sometimes interrogated under hypnosis in the hope of picking up memories not otherwise retrievable. These uncontrolled situations do not allow clear answers and are prone to biased interpretation. A negative answer in controlled experiments would obviate wild goose chases and wasted effort; a positive answer would lead to a search for the best way to employ hypnosis.

In their review of this issue, Klatzky and Erdelyi (1985) stated that none of the about two dozen studies of the effect of hypnosis on memory retrieval allowed an answer because the designs failed to rule out confounding from the criterion.

b4. Explain the predictions of threshold theory and SDT for the second-guess choice task (page 665). Is the SDT prediction sensitive to constancy of d'? Is this good or bad?

b4ANS. Threshold theory assumes that detection is a yes–no matter, occurring when and only when the signal exceeds the threshold. Further, noise alone rarely if ever exceeds the threshold. If the signal was not detected in the interval in which it occurred, it cannot be detected in some other interval because they contain only noise.

SDT allows X_N to exceed X_{SN} on some trials. In this case, the subject's first choice will be incorrect. But the correct choice may still contain some signal information that can be utilized and this will appear in correct second guesses.

No, this prediction is not sensitive to constancy of d'; it holds regardless of any exact model. This is good because it disconfirms the threshold conception with a minimum of auxiliary assumptions.

b5. Suppose subjects make two responses in a yes-no task: first a *yes* or *no*, then a rating of confidence in the correctness of the *yes* or *no*. Show how these data can provide a qualitative test between SDT and high threshold theory.

b5ANS. Under threshold theory, the confidence ratings following *no* must be independent of whether a signal was present. Nondetection means the signal was below threshold and so not perceived in any sense. Hence the *no* response must yield equal mean confidence on N and on SN trials.

Under SDT, the confidence ratings following *no* responses will on average be higher on SN than on N trials. On some SN trials, the operative X_{SN} value will be below criterion, which yields a *no*, but still higher than the average signal from the X_N distribution.

b6. You and your dog participate in an olfactory discrimination task. Sketch a graph to compare N and SN distributions of Figure 20.6 for you and for your dog.

b6ANS. Let Figure 20.6 represent your olfactory performance. Note that the horizontal axis has units of standard deviation, which are much smaller for your dog. Thus you can represent your dog by two much narrower distributions with the same means. Alternatively, you can represent your dog by shifting its SN distribution far to the right.

b7. You obtain ratings in a standard SDT task for N trials and for SN trials. You plot a histogram of each set of ratings and notice that the SN ratings have a larger variance. "Aha," you think, "here is direct evidence for the hitherto untested assumption about the relation between these two variances that is frequently made in SDT."

    a. Why must your rating procedure yield a linear scale (Section 19.3.1) for this variance conclusion to be completely certain?

    b.* How could you test the linearity assumption?

b7ANS. a. The variance is calculated from differences between individual scores and their mean. With a linear scale, the distribution of ratings for N trials will be an exact proportional image of the true distribution of sensation, and similarly for SN trials.

With a nonlinear scale, however, these numerical differences may be equal even when the underlying psychological differences are unequal; and vice versa. Granted that the SN ratings tend to come from higher on the sensation scale, the larger variance for the SN is suggestive but not certain.

b. In principle, you can establish linearity of the rating scale with an integration task, as illustrated in Figures 20.1–20.3. One such task would be to have subjects rate average signal strength for two signals presented in a row. Observed parallelism would be evidence for response linearity.

b8. Verify d' and c for these three 2×2 tables, patterned after Table 20.1.

70	30
30	70

$d' = 1.048$
$c = 0$

90	10
10	90

$d' = 2.564$
$c = 0$

70	30
50	50

$d' = .524$
$c = -.262.$

b9. Derive the "correction for guessing" implied by threshold theory for the *yes-no* experiment to obtain the "true" hit rate (Equation 4 in the text).

b9ANS. The key is that the guessing rate equals the false alarm rate, FAR, under the threshold hypothesis. The observed hits include true detections, which occur with probability s, plus guessed hits, which occur with probability $(1 - s)$ FAR. Hence

$$HR = s + (1 - s) FAR.$$

b10. In applying SDT to recognition memory in Section 20.3.3, both the "old" and "new" faces are represented as distributions. Explain how the concept of a distribution of faces of varied familiarity applies to

    a. The set of "old" faces.   b. The set of "new" faces.

    c. In what conceptual way do these familiarity distributions differ from the distributions of X_N and X_{SN} for sensory signals in Figure 20.6?

b10ANS. a,b. Each face is considered to have its own familiarity value. Although each set of faces, "old" or "new," will usually be a handy sample, we may reasonably expect the distribution of familiarity values in each set to be approximately normal, at least in many applications. The face presented on any trial, accordingly, may be treated as a random sample from some normal distribution.

c. The concept of noise for faces differs conceptually from that for sensory signals. Each face has a distinct suprathreshold identity. The same applies to many other stimulus classes studied in memory theory. Each such stimulus can ordinarily be presented only once to a given subject. The "noise" is largely real perceived differences between faces (plus a little uncertainty in familiarity values and/or criterion).

In contrast, the same sensory signal may be presented many times because the overlay of noise makes the detection process probabilistic across trials. With sensory signals, the two focal quantities are μ_N and μ_{SN}, so pooling yes-no judgments over trials is just what needs to be done, without regard to responses on individual trials. With memory stimuli, however, pooling choice responses over trials loses information that may be important.

With many memory stimuli, familiarity is a meaningful property of each individual stimulus. Measuring familiarity of each stimulus would be relevant, perhaps essential, in some memory tasks. But this cannot be done with SDT (see discussion of Thurstonian pair comparisons in Anderson, 1981, Section 5.3).

An alternative is to replace the choice responses with a continuous response. For example, subjects could be asked to rate familiarity value of individual memory stimuli. Each single response would thus yield a measure of familiarity.

Continuous response may yield only a monotone scale, but that would suffice for some purposes (Sections 7.2.3 and 19.3). In other tasks, a linear response is needed. Exercise c1 suggests a functional measurement approach to response linearity.

b11. Do you think d' can provide meaningful measures of sensory acuity for nonverbal organisms? We can't just tell a bee, "Fly to whichever color chip looks bluer" or a bloodhound: "Bark if this odor [giving sample] is present in the next whiff of air." How can we get meaningful choice responses from nonverbal organisms?

Will cross-species comparisons be meaningful?

b11ANS. Some sensory reactions are instinctive, as with pheromones, sexual odor attractants with insects. Usually, however, it would be necessary to condition the organism to use the sensory signal as a cue for reward. Continuous response measures, such as rate or time, may be superior to choice (Note 20.3.3c). Thus, visual sensitivity in bees has been studied by training them to go to the bluer of two color chips to find nectar. One interesting finding is that bees are sensitive to ultraviolet to which humans are blind, although a d' measure is not required to show this.

Meaningful comparison across species requires that the stimulus display and response mode be harmonious with the sensory and motor capabilities of each organism. Both stimulus display and response mode must be rather different for testing sensory acuity in dogs, bees, and birds. The cited training technique for bees takes advantage of the ecological importance of color cues in bee behavior. However, attentional span might contribute variability and lower d' even though attention is not properly sensory.

b12. (Not included in the text Exercises.) The following data were obtained to test the prediction of constant d' of signal detection theory. Subject's criterion was manipulated by varying signal probability; lower signal probability causes subjects to be more cautious, and set a higher criterion, thereby reducing false alarms. Theoretically, d' should not be affected by this manipulation.

Subjects received continuous background noise through earphones. On each trial within a given session, a 0.1-s, 1000-hz signal tone was presented with specified probability, known to the subject at the beginning of each session. Warning light signaled subject to get ready on each trial. Trial-cycle light was ON for the 0.1-s duration, which might or might not contain a signal. Subject pressed Yes or No button. Two separate sessions for each of five signal probabilities.

    a. What is the main trend in Hits and False Alarms and what does it mean?

    b. Get d' values for these data. (The two values in the first row are .794 and 1.570.) Plot the values of d' and interpret by visual inspection.

    c. Do Anova on d' for subject B and interpret. What statistical justification is needed to apply Anova to these data?

    d. Do you think a repeated measures Anova should be performed for the two subjects combined?

    e. Do regression analysis for subject A and interpret.

b12ANS. a. Hits and False Alarms both increase as signal probability increases. Both changes result from a more liberal criterion, as indicated in the statement of the problem.

b. Visual inspection shows a clear decline of d' for subject A, from 1.5 to 0.9. In contrast, subject B shows a near-constant d', in agreement with the fundamental formula of SDT. This difference between the two sub-

jects is perplexing. Even in this bare-bones perceptual situation, individual differences can be substantial.

c. Visual inspection gives no reason to question constancy of d'. Subject B yields $F = .65$ for probability. Of course, a linear trend test would be more appropriate, as in (e).

Any statistical test requires independence of these 10 observations. Independence seems reasonable since each observation comes from a different session.

d. I see no point to a repeated measures Anova that pools the two subjects (Exercise 11.1).

e. The regression analysis for subject A will show a statsig linear trend, using a pure error on 5 df. It solidifies the visual inspection. Notice that replication was needed to get an error term for this significance test. These data are not so reliable that visual inspection can be completely relied on.

	Subject A				Subject B	
Prob.	HR	FAR	z_{Hit}	z_{False}	HR	FAR
0.1	.37	.13	−0.332	−1.126	.47	.05
0.1	.29	.05	−0.553	−1.645	.29	.02
0.3	.53	.22	0.075	−0.772	.58	.10
0.3	.48	.19	−0.050	−0.878	.59	.16
0.5	.82	.46			.75	.34
0.5	.60	.34			.69	.37
0.7	.78	.54			.83	.53
0.7	.79	.50			.82	.47
0.9	.88	.59			.95	.74
0.9	.96	.79			.97	.86

SOURCE: (From Tanner, Swets, & Green, unpublished paper, 1956.)
I wish to thank Dave Green for making these data available.

c. Additional Exercises on Mathematical Models

c1. You wish to measure recognition familiarity for two classes of common English words for a certain type of aphasic patients in their normal state and under three experimental conditions. Application of SDT, however, faces two difficulties. Many familiarity values are above threshold, so SDT is hardly applicable. Even for the threshold choices, moreover, you would need many trials to get reliable frequencies of Hits and False Alarms.

Accordingly, you wonder whether the information integration approach discussed in the last subsection of Section 20.3.3 can be applied. You construct stimulus pairs using a 5×5 factorial design and ask subjects for two ratings of each pair: of the average familiarity and of the difference in familiarity of the two words in each pair.

    a. Following the cited text discussion, how do you use the factorial data to get familiarity values for each stimulus?

    b. It would be a lot simpler to obtain familiarity judgments of each stimulus separately. What is the advantage of getting judgments of pairs?

c1ANS. a. The marginal means of the factorial design should be a linear scale of the familiarity values, if the data exhibit parallelism. Further, these scales should agree across the two tasks.

b. One advantage of getting judgments of pairs is that these judgments can provide evidence that the subject is using the rating scale in a true linear way. Response linearity would be supported by parallelism in the two-way design. Indeed, the experiment might include ratings of the singles in concert with the pairs (one advantage of the averaging task), and these single ratings should be linearly related to the marginal means.

c2. You present the experimental plan of Exercise c1 to your research group. Someone points out that each word gets repeated presentations so its familiarity value may increase over trials.

a. What is the easiest way to test this possibility?

b. With the aphasic subjects, you decide you cannot take any chance about the objection of (a). This means you can present each word only once to each subject. How can you use the Latin square idea of Section 14.3 to get an efficient design?

c2ANS. a. The easiest way to test for a practice effect on familiarity is to compare ratings for the first presentation of a given word with those of the second. This approach could be readily extended with a Latin square to include all the data for greater sensitivity.

b. To present each stimulus just once to each subject, use a design of the Greco-Latin square type. Table 14.6, for example, would allow 16 A_jB_k stimulus pairs, of which each subject judged only four, with no stimulus repetition. It might be preferable, of course, to use a Latin square to control effects of repetition.

c3. Ratings of familiarity can provide a more sensitive measure of memory than choice data. To illustrate, suppose we select two random samples of visual scenes from a large population and give familiarity training with one fraction to get a set of "old" stimuli, the other fraction serving as "new" stimuli (balanced across subjects to play safe). Later, we obtain familiarity ratings of half of each fraction under the subject's normal state and half under a certain medication. This yields a 2 × 2, repeated measures design. For each subject, we get $(\bar{Y}_{old} - \bar{Y}_{new})$ for each condition. Any effect of condition, positive, negative, or zero, will appear in the difference between these two differences.

The rating response provides a direct measure of memory, not requiring any assumption about normal distributions. Far fewer trials are needed than with yes-no choices.

a. What assumption is involved in this comparison? How serious do you consider it?

b.* How might this assumption be tested?

c3ANS. a. The measured effect is a double difference, specifically, the interaction in the 2 × 2 design. It may thus differ from 0 because of response nonlinearity (Chapter 7).

Response nonlinearity would not seem a serious problem if $\bar{Y}_{new}$ was equal under both conditions. However, $\bar{Y}_{new}$ may well differ substantially in the drug condition, so possible nonlinearity could trouble the critical comparison.

b. One way to test response linearity is to present stimulus pairs and ask for judgments of average familiarity. Construct the stimulus pairs to include old-old, old-new, new-old and new-new pairs in a 2 × 2 design. This average judgment may be expected to obey an addition model, which implies parallelism. Observed parallelism would then support response linearity (see Parallelism Theorem in Chapter 21).

c4. Organisms may use two strategies to deal with a stimulus field. The integration strategy is to integrate the elements of the field into a unified whole, thereby utilizing all available information. Integration strategy requires substantial processing capacity, however, and a nonintegration strategy may be more cost-effective, responding in terms of just one salient element in the stimulus field.

One kind of nonintegration strategy arose with Piaget's centration hypothesis of Exercise a9. Here is a more complex centration strategy. Each individual child centers on factor A on some trials, on factor B on the other trials. You obtain continuous, numerical responses as before.

a. Show that the factorial plot, averaged over trials, will be parallel for each child (disregarding response variability), as though they added on each trial.

b. Show how to choose levels of A and B so that the distribution of response to some AB combination will differ under the nonintegration and integration hypotheses.

c4ANS. a. The proof is essentially the same as in Exercise a9.

b. Choose levels of A and B that yield quite different responses, say, high and low, respectively. Then the response to the combination will be unimodal under the integration hypothesis, bimodal with larger variance under the nonintegration hypothesis. This variance test supported the integration hypothesis (Anderson & Cuneo, 1978, p. 363).

c5. An alternative to Sternberg's additive-factor model considered in Section 11.4.4 assumes that additivity is an artifact of all-or-none responding. In this "alternative pathways model" (Roberts & Sternberg, 1993), only one process is operative on each trial. If each of two variables in a factorial design affects just one of these two processes, the data will appear additive although the two factors are not additive.

Much the same problem arose with Piaget's centration hypothesis, as noted in the previous exercise. Show how the argument in that exercise can be transposed to test the alternative pathways model.

c6. Many judgments are *comparative*: The focal stimulus is judged relative to some stimulus that serves as a comparison standard. A basic problem of processing order arises when the comparison standard has two or more dimensions. Are these separate dimensions integrated, and then the comparison is made to this integrated standard (IC order)? Or are separate comparisons made on each dimension, and then these comparisons integrated (CI order)?

The IC and CI processing orders have been empirically distinguished in certain applications, for example, in adaptation-level theory and in fairness theory. To illustrate the idea, let F denote the focal stimulus, C and D two dimensions of the comparison standard. Denote the comparison process by /, and suppose the integration follows an additive rule. The two orders may then be written

IC order: F/(C + D).

CI order: F/C + F/D.

a. How can the algebraic structure of these two models be translated into an experimental test between them?

b. In Helson's (1964) theory, the adaptation-level is an integration of all relevant stimuli in the field, and the focal stimulus is compared to it. Apparent length of the centerline in the line-box illusion of Figure 8.1 depends on some comparison of the centerline with the two flanking boxes. Show that Helson assumes the IC order. How would you test it against the CI order for the line-box illusion?

c6ANS. a. The CI order predicts that C and D will have additive effects, as shown by its algebraic structure. The IC order predicts nonadditivity with specified direction.

b. Helson assumes, as stated, that the adaptation-level is an integrated resultant of all relevant stimuli in the field. The focal stimulus is compared to this—the IC order.

To check adaptation-level theory for the line-box illusion, vary the sizes of the two boxes in a factorial design. Helson predicts nonadditivity in a specified direction, whereas the CI order predicts additivity of the two boxes.

The usefulness of this algebraic approach has been supported by success of the CI model with the line-box illusion (Clavadetscher & Anderson, 1977) and with judgments of unfairness (Anderson & Farkas, 1975).

c7. As a check on the size–weight illusion of Figure 20.3, subjects could be asked to lift two weights in succession and judge their average heaviness. Consider a 5×5 design with gram weights of 200 to 600 g for each factor. Suppose the factorial graph exhibits parallelism.

a. From the discussion of Figure 20.3, what does this factorial graph say about the psychophysical function?

b. How exactly does (a) provide a check on the size–weight study? What could this suggest about the relation between nonconscious and conscious sensation?

c. What relation may be expected between the marginal means for the row and column factors?

c7ANS. a. Just as in Figure 20.3, the vertical elevations of the row curves in the factorial graph are a visual image of the psychophysical function. A similar curve can be obtained by plotting the graph in the complementary manner. An order effect may cause these two curves to differ, although they should still be linearly equivalent.

b. Ideally, the marginal means from this weight averaging experiment are linearly equivalent to the row means from the size–weight data. In the former task, the heaviness sensation of the weights themselves is nonconscious, but conscious in the latter. Equivalence of the two would suggest that consciousness taps directly into nonconsciousness, not through some intermediate process.

c. See answer to (a).

c8. Problems of serial integration appear in every area of psychology. A plausible hypothesis in some such tasks is that the integration is a weighted sum or average of the values of the informer stimuli at successive serial positions. One example is the probability urn task, in which subjects revise their opinion about the composition of the urn as successive samples are drawn. To study serial integration task in this urn task, you present single red and white beads on successive trials, purportedly drawn at random. Subjects see only one bead at a time; at the end they make a numerical judgment of the probability that the urn contains more red than white beads.

a. What are the values of the two informers? Is it reasonable to think these values are independent of serial position?

b. You wish to get the serial weight curve for six serial positions. But this would require $2^6 = 64$ sequences, which you consider too many. How could you use fractional replication to reduce the number of sequences (see experiment of Figure 15.1)?

c. You wish to study serial position effects for sequences of 12 informers. If you use a complete serial-factor design, how many sequences are needed?

d. How can you get partial information on the serial position curve of (c) using the fractional design for 6 positions of Table 15.3 and Figure 15.1?

c8ANS. a. The values of the informers should be 1 for red beads, 0 for white. I cannot imagine why these values should depend on serial position (in contrast to weights).

b. Use the same fractional design listed in Table 15.3.

c. $2^{12} = 4096$.

d. One way to get partial information is to couple successive pairs of serial positions so they are, say, red-red or white-white. This pattern might be disguised by including irregular sequences. With this coupling, the design reduces to 2^6, which may be analyzed in the same way as in (b).

ANSWERS FOR CHAPTER 21

Exercises for Addition Model

a1. Can you average length by visual inspection? To find out, present two unequal lengths in 3 × 4 factorial design and judge average length of each of the 12 pairs. After practice trials, present the design twice to provide an error term for Anova. Do this experiment with two subjects, yourself and one other. Get verbal reports of how subjects think they did the task. For this exercise, use a 1–20 rating scale. To avoid end bias in the ratings, use *end anchors* to be called 1 and 20, respectively (Note 21.6.1a). Analyze and discuss the data.

a1ANS. The answer should include a factorial graph, with visual inspection for any apparent deviation from parallelism. Main effects are obvious and should not be reported. The overall interaction residual on 6 df is a valid test of deviations from parallelism, although it may be less sensitive than visual inspection. (The most sensitive test for nonparallelism would probably be the linear × linear component of the interaction residual, obtainable in this case by using the physical line lengths to generate the contrast coefficients. This goes beyond what is expected of students here.)

Also included should be some phenomenological discussion of task performance. Length averaging may elicit a general integration process.

The stimulus display deserves consideration. Perhaps the most informative display would place both lines at the same horizontal level, with a moderate space between them. A vertical alignment with a common endpoint would change the task into a bisection of the difference in length.

a2. In a 2 × 5 design, suppose the row stimuli have values 1 and 3, and the column stimuli have values 1, 2, 4, 8, and 16.

a. Assume an addition model, calculate the cell entries for the factorial table, and plot these values, showing that they exhibit parallelism.

b. In this same 2 × 5 design, suppose the row stimuli have values a and b, and the column stimuli have values v, w, x, y, and z. Assume an addition model and write down the entry in each cell of the factorial data table.

c. How does the factorial table for (b) reveal parallelism?

a2ANS. (c) In each column, the difference between the upper and lower entries equals $a - b$. This is constant over columns, the numerical equivalent of parallelism.

a3. Conclusion 2 of the parallelism theorem says that the marginal means for each factor are a linear scale of the values of that factor. Check this numerically with the column factor for the numerical example of Exercise a2.

a3ANS. The column means from Exercise a2 are 3, 4, 6, and 10. These columns means equal the original stimulus values + 2. Algebraically, $\psi_k = \bar{R}_{.k} + 2$, which agrees with Conclusion 2. (The zero point of the original values is not recoverable.)

a4. With an additive model, the cell entries are completely determined by the marginal means. The formula for a two-way design is

cell mean = row mean + column mean − overall mean.

Get marginal means for the data of Exercise a2 and check that this formula is correct. (As an optional exercise, derive this formula algebraically.)

a5. a. How, according to the parallelism theorem, do the data of Figure 20.3 relate to the subjective values of heaviness?

b. The psychophysical function is the function relating the subjective value of some sensory dimension to the objective, physical magnitude. How does the measurement of (a) solve the problem of determining the psychophysical function for heaviness?

c. Why go to this illusion to get subjective heaviness? Why not eliminate the illusion by concealing the weights behind a screen and ask subjects to judge heaviness of these unseen lifted weights?

a5ANS. a,b. By Conclusion 2 of the parallelism theorem, the row means from Figure 20.3 are a linear scale of the subjective heaviness of the gram weights listed by the respective curves. Plot the row means as a function of gram weight. This is the psychophysical function for heaviness, validated by the observed parallelism.

c. With judgment of unseen weights, we have no validity check on the linearity of the response measure. The need for a validity check is shown by the sharp difference between rating and magnitude estimation in Figure 21.2.

With the illusion, we have two variables and hence a potential validity check on the linearity of the response measure. This potential is actualized by finding that the illusion obeys an adding-type model. This result is one of the first, if not the first, assessment of a psychophysical function that rests on a firm measurement-theoretical foundation.

a6. In Figure 21.4:

a. Why is the parallelism essential to show that bimodality is real? b. Explain bimodality in terms of propensity to approach and avoid goals. c. Outline an experiment to test bipolarity with another stimulus class.

a7. An occasional pitfall in data analysis is to treat numbers at face value, as though they came from a proportional scale with a known zero. Usually, however, the zero point is unknown. Here is an example (Rosnow & Arms, 1968).

In a test between averaging and adding models for attractiveness of social groups, subjects made preliminary ratings of single faces on a 0–100 scale of friendliness. Then 50 was subtracted from each rating on the assumption that the midpoint of the rating scale was its true zero. These reduced ratings were used to construct paired groups, one with four faces, one with three, to yield specified differences in mean value and total value.

In one illustrative pair, the four faces in one group had values 17, 18, 17 and 16; the three faces in the other group had values 22, 19, and 22. Subjects were instructed to choose which group in each pair was more friendly.

    a. What results did the authors predict from the two models?

    b. What results are predicted by these models if the true zero of the scale in the initial face ratings is really 60?

    c. Following the logic of Figure 21.5, how could you get a valid test between averaging and adding in this task of perception of social groups?

a8. You are in an ethical dilemma in which every action, including inaction, has both positive and negative consequences. Instead of drifting with events and hoping for the best, you decide to apply utilitarian calculus and choose the action you expect to yield most good or least bad. Utilitarian calculus requires that you assign a value to each consequence of a given action, weighted as appropriate by the probability of that consequence, and add these up to get the net value of that action. You choose whichever action has highest net value.

    Assume for illustration that one specific action leads to three separate, independent consequences, with measured values of $+1$, $+4$, and -8. You assume an additive rule, which implies that the net value of this action is -3.

    Stop! Perhaps your value scale is only monotone. Perhaps the true value, ψ, follows a law of diminishing returns relative to your measured value, S.

    a. Suppose $\psi = \sqrt{|S|}$. Show that the true value of the action is positive.

    b. What is the relevance of Figure 21.2 to this particular problem?

    c. What is the relevance of Figure 21.8 to this problem?

a8ANS. a. If $\psi = \sqrt{|S|}$, the sum of true values of the three stimuli is $1 + 2 - 2.828 > 0$. Note that this task requires not merely a linear but a proportional scale (with known zero).

b. Figure 21.2 shows that two common methods for measuring value assign very different numbers to the same object. This highlights a vital missing link in utilitarian calculus, especially in its modern dress of multiattribute analysis—lack of valid theory of psychological measurement.

c. Figure 21.8 is relevant because it shows how to test validity of methods for measuring values. Doubly relevant because it shows that one of the two methods is valid.

a9. Prove Conclusion 1 of the parallelism theorem (assume errorless response). Use Equation 1 to express the entries in row 1 of the factorial design, namely, R_{1k}, in terms of the ψ values. Do the same for R_{2k} and subtract.

a10. a. Prove Conclusion 2 of the parallelism theorem (assume errorless response). Use Equations 1 to get the column means.

    b. Extend (a) to allow each level of each factor to have its own weight as well as its own value. Comment.

a10ANS. To include weights, replace ψ_{Aj} by $\omega_{Aj} \psi_{Aj}$. Weight and value are completely confounded in this product so the analysis automatically takes the unknown weights in stride. The

disadvantage is that weight cannot be separated from value.

a11. Why does Figure 21.2 imply that at least one of rating and magnitude estimation yields a nonlinear scale?

b. Exercises for Multiplication Model

b1. In a 2×4 design, suppose the row stimuli have values 1 and 3, and the column stimuli have values 2, 3, 7, and 9.

    a. Assume a multiplication model and calculate the cell entries for the factorial table. Plot these values to demonstrate the linear fan pattern, using the marginal means spacing prescribed by the linear fan theorem.

    b. Repeat (a) assuming the column stimuli have values of 2, 7, 3, and 9.

    c. In (a), the zero point constant, c_0, has been taken as 0, as though the response scale was a proportional scale. In general, however, c_0 will be unknown. To see whether the linear fan theorem still holds with only a linear scale, include c_0 and repeat (a).

b2. In Table 21.2, suppose the Coke entry of 84 is changed to 54. Show that this deviation from the multiplication model produces a nonlinear fan graph.

b3. For the size–weight experiment of Figure 5.3, plot a factorial graph to show the pattern expected from the density hypothesis. Assume plausible subjective values for size and weight. How sensitive do you think the pattern in your graph is to your particular choice of subjective values?

b3ANS. a. The density hypothesis for the size–weight illusion is a division model, Heaviness = Weight ÷ Size. It will show a linear fan, but with curves converging as size increases.

b. The density hypothesis may be qualitatively correct without being exact. If so, some sign of a fan pattern should appear in the factorial graph. The factorial graph thus provides a robust test, sensitive to the qualitative form of the density hypothesis.

b4. There are five forms of algebraic model with three variables. Four are $A + B + C$; $AB + C$; $A(B + C)$; and ABC.

    a. You are given a three-way data table and told that it follows one of the four listed forms. How can graphs be used to diagnose which one?

    b. What is the fifth form of three-factor model? How is it distinguishable from the others? (Note that $A + BC$ has the same form as $AB + C$.)

b4ANS. b. The fifth form of model is $AB + AC + BC$. Like ABC, all two-factor graphs are linear fans. Unlike ABC, the three-way interaction residual is zero.

b5. Do language quantifiers function as multipliers? This exercise asks you to test this hypothesis empirically.

    a. Get one subject, perhaps yourself, and obtain likableness judgments of persons described by adverb–adjective combinations. Use the four adverbs {*slightly, fairly, quite, extremely*} and the two adjectives {*sincere* and *insincere*} to obtain eight person descriptions. After practice (Note 21.6.1a), present two randomized replications, using a 0-20 or graphic rating.

b. Plot the data by spacing the four adverbs on the horizontal axis according to the judgments of the *sincere* person, averaged over the two replications. (This yields a perfectly straight line curve for the *sincere* person, an irregular curve for the *insincere* person.) What does the multiplication model predict about the curve for the *insincere* person?

c.* The factorial graph has one ascending curve, one descending. Their average is nearly flat, so the marginal means are unreliable estimates of the functional values of the adverbs. Yet both curves contain equivalent information about appropriate spacing of the adverbs on the horizontal. How can both be combined to get more reliable estimates?

Get these more reliable estimates, replot the data, and compare with (b).

d.* The adjectives have quantitative content in themselves; a *sincere* person is sincere to some degree. Show how your data can measure this implicit quantifier; just give the idea, using a graph.

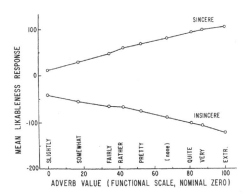

Figure 21.x. *As-if* multiplication model for adverb–adjective combinations supported by linear fan. Adverbs spaced on horizontal axis according to their functional scale values as prescribed by linear fan theorem. (From Anderson, 1974c.)

b5ANS. b. If the model holds, the judgments of *sincere* person are estimates of the values of the adverbs on a linear scale by virtue of Equations 3a,b. In principle, therefore, the judgments of *insincere* person should also follow a straight line, although irregular because of response variability. (This tactic is not a valid statistical test of goodness of fit because it does not allow for the unreliability in the judgments of the *sincere* person. Linear fan analysis provides a valid test.)

c.* Transform the descending curve into an ascending curve by subtracting the values of its points from some constant, say, the upper end of the response scale. Use the average of these two ascending curves to define the scale values of the adverbs for the graphical analysis. On this latter graph, however, plot both curves in terms of their original values. A published example based on group data is given in Figure 21.x.

d.* The implicit quantifier is implicit in the judgments of the person described by that adjective alone, without any adverb. To measure the implicit quantifier, place the judgment of *sincere* on the curve obtained from the four adverbs, represented by *none* in Figure 21.x. Drop a line down to the horizontal axis; this is the value of the implicit quantifier on the same scale as the adverbs. Any difference between the implicit values will appear as systematic differences between horizontal locations.

NOTE: Cliff (1959) claimed that adverbs do multiply adjectives, but this claim is incorrect; golden mean adjectives such as *cautious* yield a sawtooth curve, not a straight line. Cliff relied on Thurstonian scaling with choice data, which obscured this discrepancy that was readily detected with functional measurement analysis.

Even for adverb–adjective combinations that do exhibit a linear fan, the cognitive process appears to be quite different from multiplication (Anderson, 1996a, pp. 402*ff*).

b6. In Aristotle's view, fairness (justice) means that reward, or outcome O, should be proportional to work, or input I. For two persons, A and B,

$$O_A/O_B = I_A/I_B.$$

In recent times, an alternative model has been popular, based on the idea of piece-rate pay under which industrial workers get paid according to the number of pieces of work they accomplish. This piece-rate model states that

$$O_A/I_A = O_B/I_B.$$

A third model comes from information integration theory, by viewing the situation as a compromise between competing forces. The decision averaging model of Equation 2c ::

$$O_A/(O_A + O_B) = I_A/(I_A + I_B).$$

a. Intuitively, why might the comparison process be easier in Aristotle's model than in the piece-rate model?

b. Show that these three fairness models are algebraically equivalent and so cannot be distinguished empirically.

c. Each fairness model can be expanded to an unfairness model by taking the difference between the left and right sides. Aristotle's model becomes

$$U_A = \text{Unfairness to } A = O_A/O_B - I_A/I_B.$$

Consider a two-way, $O_A \times O_B$ design, with fixed values of I_A and I_B. Show how the pattern of the two-way factorial graph can distinguish between the three unfairness models.

d. All four factors were manipulated in the experiment of Figure 21.9, which shows the six two-factor graphs (see also Note 7.5.1b, page 213). Interpret these data patterns in terms of each unfairness model.

e.* (Not in text.) How might Aristotle's model be applied to analogies?

b6ANS. b. Equivalence of the first and second models can be shown by multiplying both sides of the second by I_A/O_B. To show equivalence of the first and third, invert the third, cancel the 1, and invert again.

c. For unfairness, the $O_A \times O_B$ factorial graph should be a linear fan under Aristotle's model, exhibit parallelism under the second model, and show a barrel shape under the decision averaging model.

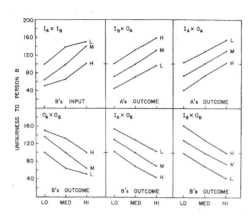

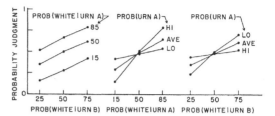

Figure 21.9. Processing structure in judgments of unfairness revealed in factorial data patterns. (After Anderson & Farkas, 1975; see further Farkas, 1991.)

Figure 21.10. Addition–multiplication model for subjective expected probability confirmed by factorial graphs. Urn A has 15, 50, or 85 white beads in 100; urn B has 25, 50, or 75 white beads in 100. Choice of urn A over urn B is made with Hi, Ave, or Lo probability; one bead is chosen at random from the chosen urn. Plotted response is judged probability that this chosen bead will be white. (From N. H. Anderson, *Foundations of Information Integration Theory*, Academic Press, 1981.)

In the experiment of Figure 21.9, all four terms in the unfairness model were manipulated. All six two-factor graphs follow the pattern predicted by the averaging model: four are parallel, and two have the shape of a slanted barrel.

Aristotle's unfairness model predicts parallelism in the same four graphs, but a linear fan pattern instead of the two slanted barrels. Qualitatively, Aristotle's model embodies the correct comparison processes, whereas the second model does not.

e.* One class of analogies has the form

A is to B as C is to D, or A/B = C/D.

An experimental test would require a task in which some or all of these four factors can be manipulated across several ordered levels, as has been done with relative quantifiers (Anderson, 1996a, p. 439, Note 5).

b7. Prove the two conclusions of the linear fan theorem.

b8. Optimal behavior follows algebraic models in many judgment–decision tasks. Such models form a base for most major approaches to judgment–decision theory. An attractive hypothesis is that cognition obeys these same models, but with subjective values of the stimulus informers. Since the parallelism and linear fan theorems allow for subjective values, they can test this hypothesis, as was done with subjective expected value in Figure 21.7.

In one judgment–decision task, the subject sees two urns, each with a specified known proportion of red and white beads. One urn is picked with specified known probability, and one bead is drawn at random from that urn. Subjects judge the probability that the bead will be white.

a. What algebraic model is implied by probability theory?

b. Figure 21.10 shows the three two-factor graphs from an experiment that varied all three factors. How well do these data support the model?

b8ANS. The probability that the bead will be white is the probability that Urn A will be chosen times the (conditional) probability of drawing a white bead from Urn A, plus same for Urn B:

Prob(White) = Prob(Urn A) × Prob(White | Urn A)

+ Prob(Urn B) × Prob(White | Urn B).

This probability model has three factors. Two factors are the two conditional probabilities; these may be represented by transparent urns with visible proportions of red and white beads. The third factor is Prob(Urn A), the probability of choosing Urn A. This probability could be represented as a white sector of a circular disc with a spinner to be spun to determine which urn would be chosen.

c. Exercises for Averaging Model

c1. Giving more informers of equal value produces a more extreme belief; this is the *set-size effect*. This set-size effect is contrary to a simple averaging model, but can be explained by assuming that subjects average an internal informer, the *initial state* or *prior belief*, together with the given informers. Including this initial state, Equation 2a becomes

$$\rho_{jk} = \frac{\omega_A \psi_{Aj} + \omega_B \psi_{Bk} + \omega_0 \psi_0}{\omega_A + \omega_B + \omega_0}.$$

a. Why is the set-size effect contrary to the averaging model without initial state?

b. Show how this explanation of the set-size effect works for the following numerical example: Each informer has value 100; prior belief has value 50; weights of 1 for each informer and for prior belief. Compare theoretical response to a single informer and a pair of informers of equal value.

c. Develop a general formula in place of the foregoing numerical example, assuming each informer has value ψ and weight ω. Set $\omega_0 = 1 - \omega$.

d. Extend the formula of (c) to allow n informers.

c2. This exercise continues Exercise 20.a5 on serial belief integration.

a. Show that prior belief has the same weight as the first serial position.

b. Use the given value parameters and the estimated weights to calculate an inferred belief for each informer sequence. Plot these as a horizontal *tree diagram*—as a set of eight curves, each with three data points for the three successive serial positions. It is common to base the tree at an assumed prior value for the initial state. The branch for sequence 101 would thus be 50, 75.00, 42.86, and 63.64.

c2ANS. a. To estimate ω_0, write out the averaging model for, say, the first sequence, 111:

$$[2 \times 100 + 3 \times 100 + 4 \times 100 + \omega_0 \times 50] \div [2 + 3 + 4 + \omega_0].$$

Set this equal to the response at the end of the 111 sequence, namely, 90.91, and solve to get $\omega_0 = 2$.

c3. Assume plausible numerical values to illustrate that the averaging model can predict the data pattern for the toy experiment of Figure 21.5.

c4. An information integration approach to signal detection theory may be illustrated with the following task of recognition memory. First, subjects receive familiarity training with a considerable number of faces, words, or other stimuli. Recognition memory is then tested by presenting a group of, say, four stimuli in a simultaneous row, some "old," some "new." The four positions in the row may be considered a 2^4 design, with "old" and "new" as the two levels of each factor. Subjects rate familiarity of the group as a whole, assumed here to be an equally weighted average of the individual stimuli.

a. How plausible do you consider the averaging model for this task?

b. How can the averaging model of (a) be tested with the indicated design?

c. Suppose equivalent informers are used at each serial position. Ideally, all four main effects are equal. Why might you nevertheless expect larger main effects at (i) later (ii) earlier postions?

d. How could this approach be used to measure effect of varied numbers of presentations of the stimuli in the original familiarity training?

e. It would be a lot simpler to obtain familiarity ratings of single stimuli rather than of a group. How could success of the averaging model show that these single stimulus ratings provide a true linear scale?

c5. To extend the decision averaging model to n alternatives, ask subjects to divide a graphic scale into n parts, each part representing the strength of one alternative. (a) How does this task relate to Equation 2c?   (b) Get a consistency check with suitably chosen groups of two and three alternatives. (c) Discuss one testable alternative to the foregoing extension to three alternatives.

c6. Individual differences can be quantified and analyzed with averaging theory. In personality theory, *response dispositions* are tendencies to evaluate another person in certain ways, independent of specific information about that person. Some people tend to view others positively, some view others negatively. Martin Kaplan (1971a,b) argued that response disposition may be considered the initial state in the averaging model. Disposition, represented by $\omega_0 - \psi_0$ and measured with the Kaplan Checklist, is thus one informer, to be averaged with given informers about some person, real or hypothetical.

Give the averaging model for this task, assuming n given informers of equal weight, ω, and equal value, ψ. Let ω_0 and ψ_0 represent weight and value for subject's disposition. Derive predictions (a)–(d) (all verified empirically).

a. For a fixed number of informer stimuli, a factorial design based on disposition and informer value will produce parallelism in the factorial plot.

b. Disposition effects will decrease as number of informers increases.

c. Disposition effects decrease as informer weight increases.

d. Subjects with positive dispositions will rate others more favorably.

e.* Outline an experimental test of the process underlying prediction (d).

f.* An alternative hypothesis is that disposition affects informer values directly. Does this hypothesis make the same four predictions?

c6ANS. The averaging model for n given informers of equal value ψ and equal weight ω may be written,

$$R(n) = \frac{n\omega\psi + \omega_0\psi_0}{n\omega + \omega_0}.$$

a. With number of informers fixed, the divisor of the averaging model is constant. The numerator, being a weighted sum of informers and disposition, predicts parallelism.

b,c. This experimental analysis is interesting because the disposition factor is individual differences that are not manipulated but observed. Kaplan tested prediction (c) in two experiments by attributing the adjectives in the descriptions to acquaintances of the person in question. In one experiment, each acquaintance was characterized by an occupation high or low in status. More weight was expected to result from a source with higher status. In another experiment, each acquaintance was characterized as being certain or uncertain that the given adjective really described the person. Both predictions succeeded. Kaplan's fine set of studies is reviewed in Anderson (1981, pp. 257-268).

An important extension is to use the averaging model itself to measure disposition. Tom Ostrom did this with prior belief in a jury trial situation (see cited reference).

e.* Kaplan measured disposition by asking each subject to check 12 adjectives they would use to describe people from the Kaplan Checklist, which contains 12 adjectives each of high, medium, and low likableness value. Disposition score was the number of high traits checked minus the number of low traits.

This question is thus complicated in that the checklist measure of disposition is similar to the response measure. Kaplan neatly resolved this difficulty by manipulating disposition experimentally by exposing subjects to a simulated radiocast that portrayed humans in a positive or negative light.

f.* Predictions (a) and (d) follow directly from the alternative hypothesis that disposition operates by influencing informer value. Under the alternative hypothesis, however, disposition effects would *increase* with number of informers, contrary to prediction (b).

A direct test between the hypotheses might seem straightforward: Just obtain judgments of the single informers. Such judgments, however, may themselves include the initial state, as with a person description based on a single informer. Almost nothing seems known about this possible complication with verbal stimuli.

c7.* Besides dispositions (previous exercise), our knowledge systems contain diverse other contents used in valuation of single stimuli, prior to integration. Can you find a way to study knowledge systems for ego defense, such as avoiding blame (Anderson, 1991j), or Freudian projection, using functional measurement to determine weight and value of stimuli?

c7.*ANS. Functional measurement can, under certain conditions, measure weight and value, not only of the initial state, but also of each single stimulus informer that is integrated into an overall response. Unlike physical measurement, psychological values and weights are not properties of the stimulus per se. Instead, they are constructed by the organism for the goal at hand, as implied by the axiom of purposiveness. This constructive process is represented by the valuation operator, **V**, in the integration diagram of Figure 21.3.

This construction of weights and values takes place through the knowledge systems of the organism. If these weights and values can be measured, they will reveal something about the nature of these knowledge systems. If the averaging model holds, it can measure these weights and values by using suitable stimulus combinations.

This approach may be considered a generalization of traditional projective tests. It may also be applicable to Freudian projection (attributing to others denied traits of the self) and other Freudian defense mechanisms (IIT considers valuation processes to be largely nonconscious). It would seem more rewarding, however, to study psychodynamics of everyday life, as with blaming and avoiding blame (Anderson, 1991j), which are more robust phenomena and more open to experimental analysis.

Other knowledge systems could of course be studied in the same way, including those that function in mood and emotion.

The usefulness of this approach is unknown. It depends on development of suitable experimental tasks (see *Personal Design*, pages 335f). For this purpose, person cognition seems a promising domain. Self-cognition has special interest, as with judgments about how the self would act under specified circumstances, or with explanations of specified, semi-hypothetical actions by the self.

Functional measurement is not a magic wand. Empirically, this approach depends on finding suitable experimental tasks. Statistically, estimation of weights and values with the AVERAGE program can run into technical difficulties. Still, this approach has unique potential for cognitive analysis of the central core of human life.

c8. In the task introduced by Schmidt and Levin (1972), subjects judge difference in likableness of two persons described by one or two trait adjectives. Let Diff_{A-B} denote the judged difference between two persons described by traits A and B, respectively; let Diff_{AX-BX} denote the judged difference between two persons described by the same A and B, plus the trait X, common to both. Assume each difference judgment equals the difference between the (implicit) judgments of the separate persons (Anderson, 1982, pp. 257f, Note 5.4h).

a. Show that an addition model predicts that $\text{Diff}_{A-B} = \text{Diff}_{AX-BX}$, but that the averaging model with equal weights makes a different prediction.

b. Show that the averaging model with equal weight implies that Diff_{AX-BX} is independent of X.

c. Suppose that more negative X traits have greater weight. What trend in Diff_{AX-BX} is predicted by the averaging model?

d. What experimental manipulation would you use to test prediction (d)?

(The paradigm introduced by Schmidt and Levin was pursued by Birnbaum (1974), who attempted to rule out an alternative interpretation in terms of nonlinear response scale. Although his attempt failed, Birnbaum nevertheless concluded that personality impressions are nonadditive, reaffirming previous studies that had used the more powerful opposite effects test, page 706; see further Anderson, 1982, pp. 257-258, Note 5.4h.)

c9. The *positive context effect* is a critical issue in the theory of information integration: If the informers in a set are interrelated in some way, the judgment of an individual informer moves toward the value of the other informers.

In one typical experiment, subjects first form an impression of a person A, described by two favorable traits and one medium trait; they then rate the likableness value of the medium trait according to "How much you like that trait of that person." These judgments are repeated for person B, described by two unfavorable traits and the same medium trait.

The data show a strong positive context effect: The medium trait is rated substantially higher in person A than in person B.

a. How can the meaning change hypothesis explain this positive context effect?

b. Alternatively, the positive context effect may be a generalized halo effect: The judgment of the medium trait is an average of its context-free value and the value of the person impression. The medium trait does not change value when the subject forms an impression of the person. Instead, the positive context effect results because the overall impression of the person acts on the post-impression judgment of the medium trait.

Some words have a broad range of meaning (e.g., *nice*), others have a narrow range (e.g., *prompt*). Several investigators used this fact to test between meaning change and halo process as explanations of the positive context effect. What was their reasoning?

c9ANS. a. The meaning change interpretation of the positive context effect is straightforward. Interaction among the adjectives as they are integrated into a unified impression produces changes in meaning and value of each to agree better with the others, thereby making a more consistent whole. The medium trait thus becomes more positive with the positive traits, more negative with the negative traits. The rating of the medium trait is simply a report of this changed meaning.

b. Words with broad range of meaning have more opportunity to change and so should change more, according to the meaning change hypothesis. This issue has been pursued by several investigators with different indexes of range of meaning. The results were almost uniformly against change of meaning. Instead, the results supported the halo interpretation, thereby buttressing the hypothesis of meaning invariance.

NOTE. The positive context effect depends on some organization among the words, such as describing a single person. No effect is obtained if the words are presented as unrelated.

c10.* (Not included in text.) Two-operation models can help study nonadditive and configural processes. Use one integration operation, such as averaging, to establish response linearity, in tandem with a second operation that is expected to show configurality. Granted a linear response, the observable data pattern for the second operation will mirror underlying process. Symbolically, the two-operation model is

$$(A \circledast B) \pm C,$$

where $\circledast$ is hypothesized to be a nonadditive operation, and $\pm$ is expected to be additive. A three-way design would be used to manipulate A, B, and C.

Two-operation logic was applied in the unfairness experiment of Figure 21.9. The difference operation predicts the parallelism observed in four of the two-factor graphs; this parallelism supports response linearity, thereby buttressing the predicted barrel shape in the other two graphs. Further, this parallelism nullifies the objection to this experiment that the data should be monotonically transformed to obtain parallelism in all six two-factor graphs (see Note 7.5.1b, page 213).

Two-operation logic seems useful to study functional inconsistency and redundancy. For example, A and B could be trait adjectives that describe one person, with some $A_j B_k$ pairs inconsistent or redundant, and others expected not to interact; C could be a pair of trait adjectives that describe a second person. Subjects could judge the difference in, or average of, likableness of the two persons, which involves preliminary integration of the A and B adjectives, $A \circledast B$.

a. Why is it plausible to expect an adding-type model for the judgment of the average or difference in likableness of the two persons?

b. What shapes of factorial graphs are predicted if the hypothesized adding-type and configural integrations hold?

c. Construct stimuli for this design, make up hypothetical data, plot and interpret the factorial graph. Alternatively, apply two-operation logic to a task of interest to you.

d. The integration of the two adjectives of the C factor may also be configural. Why does this not trouble the two-operation logic?

c10*ANS. a. Adjectives that describe a single person might well interact from inconsistency or redundancy because they are aspects of a single entity that must cohere in some manner. But adjectives that describe two unrelated persons are not under such an integration constraint.

c. The inconsistency/redundancy task was used for simplicity in the question, although this particular experiment has not been performed. For other experimental applications of two-operation logic, see index entries in Anderson (1982).

ANSWERS FOR CHAPTER 0

1. In the numerical example of Section 0.1.3, suppose 9 is changed to 14.

   a. By visual inspection of Equation 1a, without actual calculation, do you think the variance will increase, decrease, or stay the same? Why?

   b. Use Equations 1 with hand calculation to show that $s = 3.536$.

1ANS. a. If 9 is changed to 14 in the numerical example, the range of the numbers increases; hence the deviations from the mean must increase. Since the variance is an average of the squared deviations from the mean, it must also increase. (The change from 9 to 14 facilitates hand calculation because the mean remains an integer.)

2. Given the information of Figure 0.1 about heights of adult U.S. women, guess the mean and standard deviation for heights of adult U.S. men.

2ANS. The sense of this exercise is to use whatever background information one has about relative heights of women and men in conjunction with the mean and standard deviation given for women's heights to get intuitive estimates for men. Although the correct answers are not important for this exercise, they are approximately 69 and 3.0 inches for mean and standard deviation. The larger mean is accompanied by a larger standard deviation, as is not infrequent.

3. Consider the half-width of the confidence interval, $t*s / \sqrt{n}$.

   a. As n increases, other things being equal, what happens to the half-width?

   b. Intuitively, why does your answer to (a) make sense?

   c. As s decreases, other things being equal, what happens to the half-width?

   d. Intuitively, why does your answer to (c) make sense?

   e. For your present experiment, what can you do to decrease s?

3ANS. a. Larger n yields a larger denominator in the half-width formula, which decreases the half-width. (Other things are not quite equal as the Exercise assumes; as n increases, so do the df, so $t*$ decreases.)

b. A larger sample contains more information and so should localize the population mean more precisely.

c. As s decreases, the half-width obviously also decreases.

d. Intuitively, (c) makes sense because lower variability means lower ''likely error'' about the location of the population mean.

e. One way to reduce variability, at least with humans, is to develop clear instructions so all subjects understand the task. Include checks on each subject's understanding as part of the initial instruction period. See further Sections 4.3.4 and 14.1.2.

4. In the numerical example of confidence interval in Section 0.1.3, suppose the numbers are difference scores, $Y_{Ei} - Y_{Ci}$, each subject in both treatments.

   a. Can we reasonably conclude treatment E is superior to C?

   b. Get the t ratio for the mean of this sample. Interpret it.

   c. Which do you prefer, confidence interval or t test? Why?

4ANS. a. Since 0 lies outside the 95% confidence interval (from 5.03 to 8.97), we may reasonably believe that the true mean difference for the population is not 0. And hence that E is superior to C.

b. By Equation 3, the t ratio for this mean difference is

$$t = 7 \div 1.581 / \sqrt{5} = 9.90.$$

This is greater that the criterial $t*$ of 2.78, and so is statsig.

c. I prefer the confidence interval because it conceptualizes—and quantifies—the mean as an interval of uncertainty And it yields a significance test to boot.

5. Under *Confidence Interval as Significance Test* in Section 0.1.3, explain in your own words the meaning of:

   a. ''Therefore, if 0 lies outside this interval, we have 95% confidence that 0 cannot be the population mean.''

   b. ''In other words, we have 95% confidence that the two treatments differ in effectiveness.''

6. You test E versus C with the same subjects in both conditions, and get difference scores of 0, 1, 2, 3, and 4 for your five subjects.

   a. What is the similarity between this sample and that in the *Numerical Example* in the text (page 787).

   b. On the basis of this similarity, guess s for this sample.

   c. What principle underlies your guess in (b)?

   d. On the basis of this similarity, find the 95% confidence interval for the mean for a significance test.

   e. Get t ratio for the mean (Equation 3) for a significance test.

   f. Which do you prefer, confidence interval or t test?

6ANS. a. These numbers are the same as those in the text, except that 5 is subtracted.

b. An obvious blind guess is that the standard deviation is the same for both. And this seems sensible because the *deviations* from the mean must be the same for both samples. Hence, by Equation 1a, both must indeed have the same standard deviation.

c. The principle underlying (a) is that adding a constant to all the numbers in a sample has no effect on the standard deviation. The statistical rationale is that adding a constant to all the numbers does not affect the differences among them—that is, the deviations from the mean. Since the standard deviation is calculated from the deviations from the mean (Equations 1), it also must be unchanged.

d. By Expression 2, the width of the 95% confidence interval must be the same for both sample means because both have the same n and the same standard deviation. Hence the 95% confidence interval must be from .03 to 3.97, so the E condition is statsig better (just barely).

e. The t ratio is 2.83, which is statsig (barely).

7. Use Equation 1a to calculate with pencil and paper the variances for these four samples with $n = 2$. Also get the standard deviations using Equation 1b.

    a. (1, 3). b. {1, 5}. c. {1, 7}. d. {1, 9}.   .br [Variance for (d) is 32.]

    e. What progressions do these four cases exhibit?

    f. Use these progressions to predict mean and standard deviation for the next two samples in the progression.

    g. Do (f) for the variance.

7ANS. a-d. The variances are 2, 8, 18, and 32, in order. The corresponding standard deviations are the square roots: 1.4142, 2.8284, 4.2426, and 5.6568.

e,f. The higher number in the sample increases by 2 each time, hence the mean increases by 1, showing a linear progression. The next two means should be 6 and 7. The standard deviation increases by 1.4142 each time, also showing a linear progression, so the next two standard deviations should be 7.0710 and 8.4852 (with rounding error).

g. The variance shows an accelerated increase; the successive increases are 6, 10, and 14, so the next two variances should be 50 and 72.

8. Length and weight of normal term, newborn males in the U.S. follow normal distributions with means of approximately 19.5 inches and 7.1 pounds, and standard deviations of approximately .89 inches and 1.08 pounds.

    a. Find the range of lengths that includes 68% of cases.

    b. Find the range of weights that includes 95% of cases.

9. Consider two samples: {1, 3} and {1, 2, 3 }. By visual inspection of Equation 1a, say which will have smaller variance. Or will both be equal?

9ANS. The average difference between the sample numbers is smaller for the second sample; so it should have lower variance.

    Quantitatively, Equation 1a says the variance is an average of the squared differences between the sample numbers and the sample mean. The mean is 2 for both samples. The squared differences for the two numbers in the first sample are 1 and 1. The squared differences for the three numbers in the second sample are 1, 0, and 1. The sum of squared differences is the same for both samples, but we must sum and divide by sample size $- 1$. This gives 2 for the first sample, 1 for the second.

10. Under *Law of Averages*:

    a. Verify that the standard deviation of the proportion of heads is .10 and .05 for samples of size 25 and 100, respectively.

    b. Verify that the standard deviation of the number of heads is 2.5 and 5 for samples of size 25 and 100, respectively.

11. Pollsters can get more accurate results with larger samples, but larger samples are more expensive. In the numerical example at the end of Section 0.1.4, how large a sample is needed to cut the margin of error in half? What does this tell you?

11ANS. To halve the margin of error from .02 to .01, solve $\frac{1}{2} \div \sqrt{n} = .01$ to get $n = 2500$. This says that *halving* the error requires *quadrupling* the sample size. This illustrates the square

root law of sample size, which holds in general.

12. You see a poll on TV saying that 56% of the voters favor the bond issue for wildlife conservation in your state, with a margin of error of 4%.

    a. How large was the poll?

    b. The bond issue needs a simple majority to pass. Taking this poll at face value, is the bond issue likely to pass?

    c. What statistical assumption underlies the validity of this poll result?

    d.* How can the statistical assumption of (c) be satisfied empirically?

    e. What clue suggests the poll may not be too trustworthy?

    f. What empirical assumptions underlie the validity of this poll result?

12ANS. a. Since $p = .56$, we solve $\sqrt{.56 \times .44} \div \sqrt{n} = .04$ to get $n = 154$.

b. Since the reported .56 is greater than .50, at face value it implies the bond issue is more likely than not to pass. No further statistical analysis is needed to tell us that.

    Since the .56 is more than one margin of error above .50, the bond issue has a fair chance to pass. At the same time, the 95% confidence interval, which is approximately ± 2 margins of error, extends below the 50% level to .48 so confidence in passage is well under 95%.

c. The statistical assumption is that these persons polled are a random sample from those who will vote.

d.* To get a random, or almost random, sample of people is a minor science, extensively developed by polling and survey organizations. Among the complications are that some people are difficult to contact, some say they don't know how they will vote, and some later change their minds.

e. The small size of this sample suggests it is an amateur job, not a random sample, and not worth much except to suggest that the bond issue may be a close call.

f. The statistical assumption of (c) is also an empirical assumption. Another is that pollees report correctly how they will vote.

13. a. Verify that a random number between 00 and 99, inclusive, has probability .18 of having exactly one digit $=$ to 2.

    b. Below Equation 7, verify "Counting shows that (A AND B) consists of 13 numbers."

    c. Show "Rewriting the conditional rule of Equation 7" yields Equation 8.

14. *Figures Always Lie* in Section 0.3.2 considers two ways in which data can lie. Give two other ways. (See also Note 0.3.2a).

14ANS. One common, effective way to lie with figures is to select special cases that illustrate the point you wish to make. Do you want to show how bad some program is? Select one or two of the failures. Or do you wish to make the program look good? If so, select one or two of the successes. The gloomy fact is that many people are more impressed by one selected dramatic case than by a table with any number of representative cases.

Another common way to lie with figures is with a graph. Stretch it or compress it to force it to say what you want to say. Do you want to show how stable some phenomenon has been over time? Take each unit on the vertical axis to represent a large amount; then the curve of data as a function of time will stay practically flat. Or do you want to show how unstable the phenomenon is? If so, do the opposite to expand the graph so the data points bounce wildly up and down.

"Crazy Radar Mechanics," from Wallis and Roberts (1956, p. 82) is a neat example. One student in their course had been responsible for repair and maintenance of aircraft radar in the Mediterranean Theater in World War II. He was authorized to have 40 to 50 radar mechanics, which were badly needed. However, he was only able to obtain seven; repeated requests and complaints accomplished nothing. One month, one of his seven men suffered mental breakdown from overwork. Seizing this opportunity, he reported "Over fourteen percent (14%) of the radar mechanics went crazy last month due to overwork." Almost immediately, he received 35 more radar mechanics.

15. The OR rule of Equation 6 is often illustrated graphically. Draw a square to represent the set of all balls in the Probability Urn. Inside the square, draw two possibly overlapping circles to represent the balls in A and B, respectively. Let the area in circle A be proportional to the number of balls in A, and similarly for B. Show that adding areas can illustrate the two listed equations for the OR rule.

15ANS. (A OR B) refers to those balls in A plus whatever additional balls are in B; duplicates count only once. In terms of areas, this equals the area of circle A plus all additional area in circle B that is not in A. The area common to A and B represents (A AND B). Hence we add area of A to area of B and subtract the common area to verify Equation 6a.

If A and B are mutually exclusive, they have no common area, verifying Equation 6b.

16. Surgeons were told that the annual rate for complications following a certain surgical procedure was 20%. For the current year, now half over, the rate was 14%. They were asked to predict the rate for the rest of the year.

　　a. What dumb mistake did the surgeons make? And why?

　　b. What is *your* prediction? And why?

(Note: This is a published study, but I have not relocated it.)

16ANS. a. The surgeons predicted the rate for the second half of the year would be 26%. They applied the "law of averages," so the lower than average frequency in the first six months has to be balanced by a higher than average frequency in the second six.

I used "dumb" as a clue to the answer, not as a slight on surgeons. Their mistake is common, even among those with some statistical sophistication.

Actually, their mistake represents a cognitive achievement that may repay research. Among other things, it reveals some understanding of the "law of averages," which, even though not correct, represents a substantial knowledge system.

b. Two predictions are reasonable, namely, 20% and less than 20%. The 20% prediction assumes that the 14% figure is just a chance fluctuation, and that the past 20% rate will apply to the second six months. It is possible, however, that the 14% rate

represents a real reduction due to improvements in treatment or hospital conditions, in which case the rate should continue lower in the second six months.

17. Granted that a coin "has no memory," it is logically impossible for it to equalize the *numbers* of heads and tails, as noted in the text. How then can it equalize the *proportions* of heads and tails? You may be able to discover the principle by considering the following example. You toss a fair coin 300 times and observe 200 heads, 100 tails—100 more heads than tails.

You decide to toss 300 more times. How many heads and tails should you expect in these additional 300 tosses? What should you expect to happen to (i) the difference in number of heads and to (ii) the proportion of heads?

17ANS. Since the coin has no memory for the first 300 tosses, you expect the second 300 tosses will yield 150 heads, 150 tails, on average. For all 600 trials, the discrepancy in *number* of heads then remains 100, exactly the same. But the discrepancy in *proportion* of heads is reduced from .667 to .583.

Simple statistical considerations thus show that the "law of averages" has a precise meaning: The sample *proportion* approaches the population proportion as sample size increases. But it does this, not by compensating for past excesses, but by swamping them in the average. Similarly for sample means.

18. How base rate can fool uneducated intuition is illustrated in this example from Christensen–Szalanski and Beach (1982).

> In a city of 100,000 people, there are 7,000 people who have contracted disease K. A test for disease K is positive in 80% of the people who have the disease and negative in 80% of the people without the disease. The test is given to all the people in the city. In this city, what is the probability that a person with a positive test has disease K?

a. The modal answer of the subjects was .80. Intuitively, do you think this is too low, too high, or about right?

b. Calculate the number of those with disease K who will test positive.

c. Calculate the number of those without disease K who will test positive.

d. From these two calculations, find the probability that a person who tests positive has disease K.

e. What is the moral of this example?

18ANS. b,c. Of the 7000 with disease K, 5600 will test positive; of the 93,000 without disease K, 18,600 will test positive.

d. Hence the probability that person with a positive test will have disease K is 5600/(5600 + 18,600) = .23.

e. The moral is that the apparent efficacy of any diagnostics procedure can be seriously misleading. The true efficacy must allow for the base rate. In this exercise, the apparent efficacy may seem to be 80%, but the true efficacy is only 23%, certainly better than 7% but still low.

Printed and bound by CPI Group (UK) Ltd, Croydon, CR0 4YY

17/10/2024

01775694-0018